THE CALIFORNIA QUAIL

A JOINT CONTRIBUTION OF
CALIFORNIA ACADEMY OF SCIENCES
MUSEUM OF VERTEBRATE ZOOLOGY
DEPARTMENT OF FORESTRY AND CONSERVATION,
UNIVERSITY OF CALIFORNIA, BERKELEY

G.M. Christman

THE CALIFORNIA QUAIL

A. STARKER LEOPOLD

UNIVERSITY OF CALIFORNIA PRESS

BERKELEY · LOS ANGELES · LONDON

UNIVERSITY OF CALIFORNIA PRESS
BERKELEY AND LOS ANGELES, CALIFORNIA
UNIVERSITY OF CALIFORNIA PRESS, LTD.
LONDON, ENGLAND

ISBN 0-520-03362-0
LIBRARY OF CONGRESS CATALOG CARD NUMBER: 76-48003
PRINTED IN THE UNITED STATES OF AMERICA

DESIGNED BY DAVE COMSTOCK
ILLUSTRATED BY GENE M. CHRISTMAN

1 2 3 4 5 6 7 8 9

CONTENTS

PART II. NATURAL HISTORY

FIGURES AND TABLES

FIGURES

TABLES

PREFACE

This volume is the cumulative product of the work of many people. I first became interested in the ecology of arid-land quails while engaged in a survey of the wildlife of Mexico in the period 1944 to 1946. When I joined the faculty of the University of California in 1946, I encouraged one of my first graduate students, Richard Genelly, to initiate a study of the California Quail in the hills east of Berkeley. Genelly completed an admirable Ph.D. dissertation on the quail, but, as is often the case, he raised as many questions as he answered. Genelly was followed sequentially by five additional Ph.D.'s, each of whom contributed substantial knowledge of quail ecology and natural history. They were: Ralph Raitt, Victor Lewin, William Francis, Richard Jones, and Richard Fletcher. Yet there still remained unresolved enigmas of the natural controls over population numbers in this fascinating species.

Concurrently with our research activities, I had the good fortune to follow year by year the evolution of a practical and highly successful management program for California Quail on the ranch of Ian McMillan of Shandon, San Luis Obispo County. Starting from scratch, with no quail whatsoever on his property, McMillan manipulated the habitat until he had built up a population that today fluctuates between one and two birds per acre on his land—a truly remarkable accomplishment. Much of what I know about quail management was taught to me by "Ike."

More recently, I have had the opportunity to go afield with Ray Conway of Grass Valley and to see on the ground the highly successful program of quail management that he has instituted on properties between Grass Valley and Marysville. In the oak timberlands of the foothills bordering the Sacramento Valley on the east, Conway has experimented with bulldozing openings in the thickets of oak saplings created by earlier wildfires, permitting the growth of shrubs and forbs that supply excellent quail habitat where few birds could exist before. Conway's management program has been fully as successful as McMillan's, though it deals with an entirely different set of problems. He likewise has developed quail densities of two birds per acre on a sustained basis, while at the same time deriving substantial income from cattle grazing on the improved pasturelands. I am equally indebted to Ray Conway for what he taught me about quail management in the timber zone.

In 1970, Peter McBean of San Francisco expressed an interest in subsidizing through the California Academy of Sciences the preparation of a book on the California Quail and its management. McBean is himself an avid quail hunter, and he owns ranch property in southern California where he would like to increase quail numbers. But, as he pointed out, there is no available guide to management of the California Quail, the 1945 pamphlet on the subject written by Emlen and Glading being long out of print. McBean's suggestion and generous financial support led to the preparation of this volume, the intent of which is to assemble in one set of covers all that is known to date about the ecology, natural history, and management of the species.

With support from the McBean fund in the California Academy of Sciences, yet one more graduate student, Michael Erwin, was assigned to study the quail population in the vicinity of McMillan's ranch near Shandon and to pull together loose ends of quail biology in relation to fluctuations in rainfall and to the ongoing management program underway on McMillan's property and that of some of his neighbors.

The initial idea of the book was to make it a collaborative effort of McMillan, Erwin, and myself. It soon became clear, however, that Erwin was completely preoccupied with his thesis project and would have little or no time to devote to book writing. During the spring and summer of 1974, I prepared a rough draft of the volume which was scheduled for co-authorship by McMillan and myself. However, some serious points of disagreement arose concerning our differing philosophies of land management, and these became more crucial when I prepared my version of a second and subfinal draft in the spring of 1975. Finally, to the sincere regret of both of us, McMillan concluded that our differences were insurmountable and he withdrew from co-authorship. A primary point at issue concerned the propriety of creating openings in chaparral or timber through prescribed burning or by mechanical means—management techniques to which I fully subscribe but which McMillan feels are potentially deleterious to the soil and to the basic integrity of woodland resources. With a sense of sorrow, I accepted McMillan's decision to disassociate himself from sponsorship or authorship of the book. Nevertheless, I wish to acknowledge the many constructive ideas and suggestions that McMillan contributed to the text. The book has been greatly enriched by his keen and perceptive observations of quail ecology.

The appendices to the volume represent the independent contributions of three scholars of quail history and behavior. Karen Nissen, a graduate student in Anthropology at the University of California, Berkeley, compiled the historical record of aboriginal use of quail by the California Indians. Bruce Browning, a Wildlife Biologist in the Food Habits Section, Wildlife Investigations Laboratory of the California Department of Fish and Game, assembled all the quail food habits data accumulated over the years in the Wildlife Investigations Laboratory. And Michael Erwin's MS. thesis in the Department of Forestry and Conservation, University of California, Berkeley, on the effect of rainfall on quail reproduction at McMillan's ranch presents the physiological and behavioral differences observed in two years of contrasting precipitation.

I am deeply grateful to the patient friends and critics who reviewed the manuscript and offered many helpful suggestions for its improvement: B. M. Browning, E. Callenbach, R. Conway, M. J. Erwin, C. M. Ferrel, D. L. Fox, W. J. Francis, S. Gallizioli, R. Genelly, R. J. Gutierrez, H. T. Harper, V. Lewin, I. McMillan, R. J. Raitt, and R. Teague.

Gene M. Christman's illustrations have added greatly to the attractiveness of the volume. My faithful secretary, Nobu Asami, cheerfully typed and retyped countless pages in the tedious process of getting the manuscript ready for the Press. And Sandra Martin and Judy Sheppard were most helpful in the final stages of proofreading and indexing.

Lastly, I wish to acknowledge the major contributions made by pioneers in the study of California Quail ecology, whose publications supplied much of the factual basis of this book. Lowell Sumner, John Emlen, and Ben Glading particularly deserve mention, but the extensive bibliography attests to the work of many other contributing biologists.

My primary hope, which I know I share with Peter McBean, Ian McMillan, and Ray Conway, is that this volume will serve as a stimulus and a guide to the preservation and management of the California Quail—one of the finest game birds of North America.

A. Starker Leopold

Berkeley
February 9, 1977

PROLOGUE

There were no quail on my father's homestead in McMillan Canyon along the western border of the San Joaquin Valley. For miles around, the countryside was open grassland or dry-farmed cropland with no shrub cover for quail.

The first quail I ever saw were bagged and brought home by my older brother in an all-day hunting trip to a distant patch of shrubland where a big covey had its fall and winter territory.

The impressions of that bag of game remain indelible. Meat for the homestead table often included game, but the quail were something new and special. With other younger members of the family I was assigned the chore of skinning and cleaning the birds. This was done after the evening meal as we gathered for warmth near the big kitchen stove. It was a raw, cold evening in late fall. A strong northeast wind was blustering outside. The homestead cabin and lean-to were of typical board-and-batten construction, and for added shelter a covering of paper was loosely pasted on the inside walls. Enough wind came through the boards to make a humming sound in the wallpaper. Occasionally the kerosene lamp on the kitchen table would flicker.

The first bird I drew from the bag was a plump cock. I marveled at the fanciful "top-knot" and the equally handsome plumage. The shrub cover where the quail had been bagged was mainly California sage, and faintly but unforgettably the birds still carried the aromatic scent of sage. There was also the rich smell of the warm innards. When the cleaning was completed, a bird for each young worker was salted and laid at the front edge of the glowing oak coals in the big wood stove. Quickly the fresh meat was singed and charred on the surface and when about half-cooked the birds were handed around for the rewarding feast.

Here the youthful impressions were particularly vivid. Under the charred crust the tender, juicy meat had a flavor that seemed to epitomize the entire experience. The story of the hunt and of finding the big covey in the isolated patch of sage, the skillful wing shooting, and the long horseback ride home all seemed blended in the feast. In family talks we had heard of the fabled "fat of the land." Here it was—the fat of a rich, new fabulous land, the taste enhanced by the odor of charred flesh, the warmth of the stove, and the sound of the raw wind howling outside.

The homestead cabin has vanished, and the family is grown and scattered. Although the isolated patch of sage still exists as winter habitat for quail, only a small remnant now remains of the big covey. The surrounding landscape has changed profoundly, with most of the changes adversely affecting quail. The entire ecosystem has been altered by decades of exploitation.

The avowed purpose of this book is to advocate a program of survival for the California Quail. Secondly, but significantly, it suggests that there is profound social value in human experience that generates a feeling for the land and its productivity. The California Quail remains in my mind a symbol of the productivity of the San Joaquin landscape—the ''fat of the land.''

Ian McMillan

THE CALIFORNIA QUAIL

PART 1:
THE BIRD AND ITS HISTORY

GMC 77

1
THE SPECIES AND ITS DISTRIBUTION

CALIFORNIA'S STATE BIRD

Of all the birds native to the state of California, none is more universally enjoyed and appreciated than the California Quail. The handsome plumage, pert demeanor, and melodious calls are appealing to everyone fortunate enough to know the species, and it is not surprising that in 1931 the State Legislature by unanimous acclamation declared the California Quail to be the State Bird of California.

Long before the Spanish occupation of California, the aboriginal inhabitants valued the quail as an important food resource. The Indians hunted and trapped quail for food and used the plumage to decorate their baskets and ceremonial clothing. During the era of land settlement, the bird became an important item of commerce, and millions were trapped or shot for the San Francisco market. With the subsequent dawning of the conservation era, commercialization was terminated but the species has continued to be utilized for sport hunting under rigid protective laws. Today the esthetic appeal of the California Quail is one of its great social values, the presence of a covey in an isolated canyon or in the immediate suburbs of a city adding a touch of interest and natural beauty to which hunters and non-hunters alike must respond.

A bird of such importance and universal appeal quite naturally attracted the early attention of naturalists and wildlife biologists. Over the years the California Quail has been observed and studied by scores of competent

Figure 1. Portrait of a male California Quail. C. W. Schwartz photo.

scientists who have published their findings in technical books and journals. Few American bird species have been more intensively investigated, with the possible exception of the closely related Bobwhite Quail in the eastern United States and some of the commonest garden songbirds. When the extensive literature on the California Quail is studied and distilled, two truisms emerge which constitute central themes of this volume:

(1) First, the local status and welfare of a quail population will be a direct function of the quality of habitat available for quail occupancy. Any area that supplies the necessary food, cover, and water required by the bird

for survival can support a quail population. Conversely, in the absence of proper habitat the bird cannot exist irrespective of protective laws or benevolent intentions.

(2) Secondly, in many situations it is possible to restore quail habitat where it has been destroyed or depleted by past land abuses. Generally speaking, the measures required to rehabilitate quail range are consonant with good soil and water conservation practice. The old adage that "good land use is good wildlife management" applies very nicely to the foothill ranges of the California Quail.

In the chapters which follow, I will attempt to summarize the accumulated knowledge about this bird—its distribution, habitat relations, natural history, and management.

NATIVE RANGE OF THE SPECIES AND ITS CLOSE RELATIVES

The California Quail, *Lophortyx californicus,* is the west coast representative of a group of closely related quails that range through the arid and semi-arid portions of southwestern North America. The species occupies virtually all of Baja California and California, except the Colorado and eastern Mojave deserts and the higher reaches of the Sierra Nevada and Cascades. Its native range includes a small portion of western Nevada and the southern tier of counties in Oregon. The species has been widely transplanted throughout the west and now occurs over most of Washington and Oregon, and in scattered localities of Idaho, southern British Columbia, Utah, and northern Nevada (Fig. 2).

The closely related Gambel Quail, *Lophortyx gambelii,* occupies contiguous range in the deserts of southeastern California, southern Nevada, Arizona, Sonora, the extreme northeastern corner of Baja California, and deserts to the east and south. Overlapping the range of Gambel Quail, and extending farther east and south is the Scaled Quail, *Callipepla squamata.* The Bobwhite, *Colinus virginianus,* overlaps the range of the Gambel Quail in southern Arizona and Sonora. The Mountain Quail, *Oreortyx pictus,* occurs widely through California, from northern Baja California into Oregon. Its range, like that of the California Quail, has been extended to the north and east by transplant. The Montezuma Quail, *Cyrtonyx montezumae,* is resident in portions of the pine-oak uplands of Arizona, New Mexico, Texas, and the central plateau of Mexico. There are, in all, six species of native quails in the western United States.

Some of these species are very similar in size, coloration, and general habits. For example, the California and Gambel quails are so alike as to be easily confused in the field. On the other hand, some species whose ranges adjoin or overlap are very different in both appearance and in habits—consider the Gambel Quail and Masked Bobwhite, both occurring originally

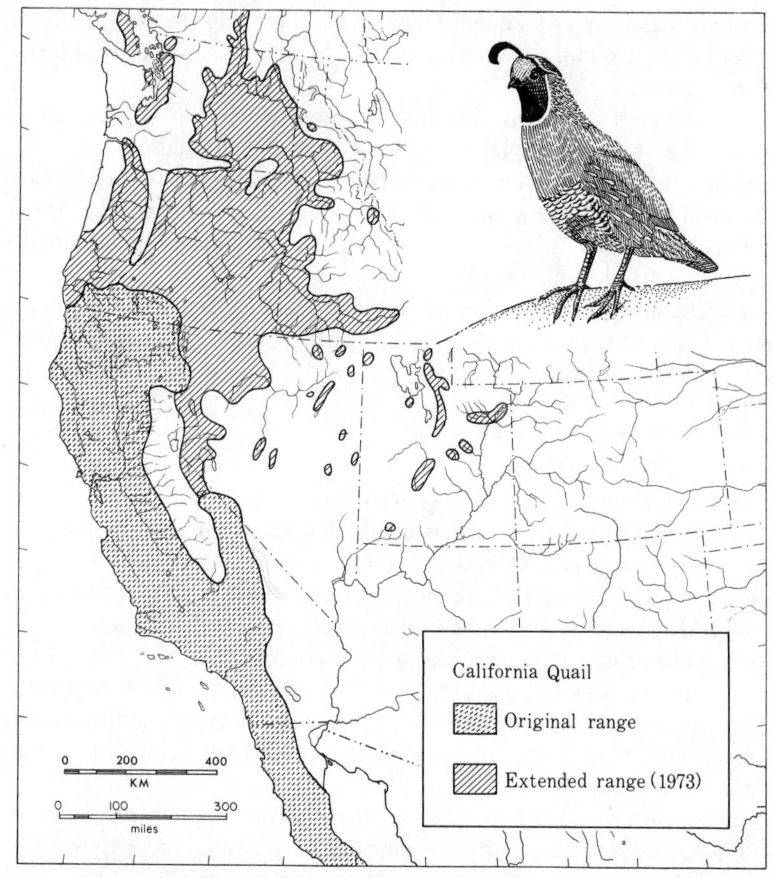

Figure 2. Original and present range of the California Quail in North America, north of Baja California. For range in Baja, see Fig. 4.

in southern Arizona (the Bobwhite is now locally extinct). The question arises, how did six species of quail come to evolve in the Southwest and what is their relationship to one another?

All the quails are classed as members of the large family Phasianidae, which according to Van Tyne and Berger (1976) embraces 174 species distributed in all continents of the world. The particular group of quails of concern to us here would seem to fall into four rather distinct groups representing independent phylogenetic lines. They are:

Lophortyx quails	*Lophortyx californicus*	California Quail
	Lophortyx gambelii	Gambel Quail
	Callipepla squamata	Scaled Quail
Oreortyx quail	*Oreortyx pictus*	Mountain Quail
Colinus quail	*Colinus virginianus*	Bobwhite Quail
Cyrtonyx quail	*Cyrtonyx montezumae*	Montezuma Quail

The group of Lophortyx species are obviously closely related, and in fact interspecific hybrids are known to occur (though rarely) between California and Gamble Quail and between Gambel and Scaled Quail. On the basis of this hybridization, Johnsgard (1973) lumps the genera *Lophortyx* and *Callipepla* into the single genus *Callipepla,* but this proposal has not yet been accepted by the Nomenclature Committee of the American Ornithologists' Union, so I shall use the traditional generic names given above.

The other three species represent quite distinct genetic lines with little in common either in habits or in general appearance. Presumably these four blood lines derived independently from some early common ancestor, and all persisted to fill different niches in this part of the North American continent.

Each species now claims as its own range a given geographic area with associated attributes of climate and vegetation. Each species, we say, is adapted to its native habitat. All individuals of the species are capable of breeding among themselves, but do not breed freely with members of other species.

In some cases, the ranges of western quails are allopatric (Fig. 3). That is to say they adjoin with a minimum of overlap. Thus the Gambel Quail supplants the California Quail along a fairly clear line extending from the Panamint Range in Inyo County to the Santa Rosa Mountains in Riverside County. To the west of this line only the California Quail occurs, whereas to the east the Gambel Quail is the sole occupant of appropriate desert washes. Along the border, the two species occur together, as for example at San Gorgonio Pass. There occasional hybridization takes place, attesting to the close relationship of the species. What, then, is the nature of the adaptation that fits each of these species to its particular native range?

By and large, we know very little about the specifics of range adaptation. Grinnell et al. (1918:540) states: "Efforts made to introduce the Desert Quail (Gambel) into northern California have met with failure. Belding (1890, p. 8) records the fact that although a number were once liberated near Folsom, Sacramento County, they all soon disappeared. A covey, numbering originally more than a hundred, kept on the State Game Farm at Hayward, slowly died off until not one was left. The bird seems unable to stand any departure from the warmth and dryness of its native desert territory." By virtue of this kind of trial-and-error evidence, plus the empirical fact that the two species maintain distinct ranges, we are led to accept the existence of adaptive differences, even though we are largely ignorant of their nature and sharpness.

A recent paper by Henderson (1971) gives some clues as to the nature of the specific adaptations of different quails to local environments. Gambel and Scaled Quails were tested in a laboratory environment to measure resistance to high temperature and deprivation of water. Evaporative water loss rates were similar at 25, 30, and 35°C. However, at temperatures of 40

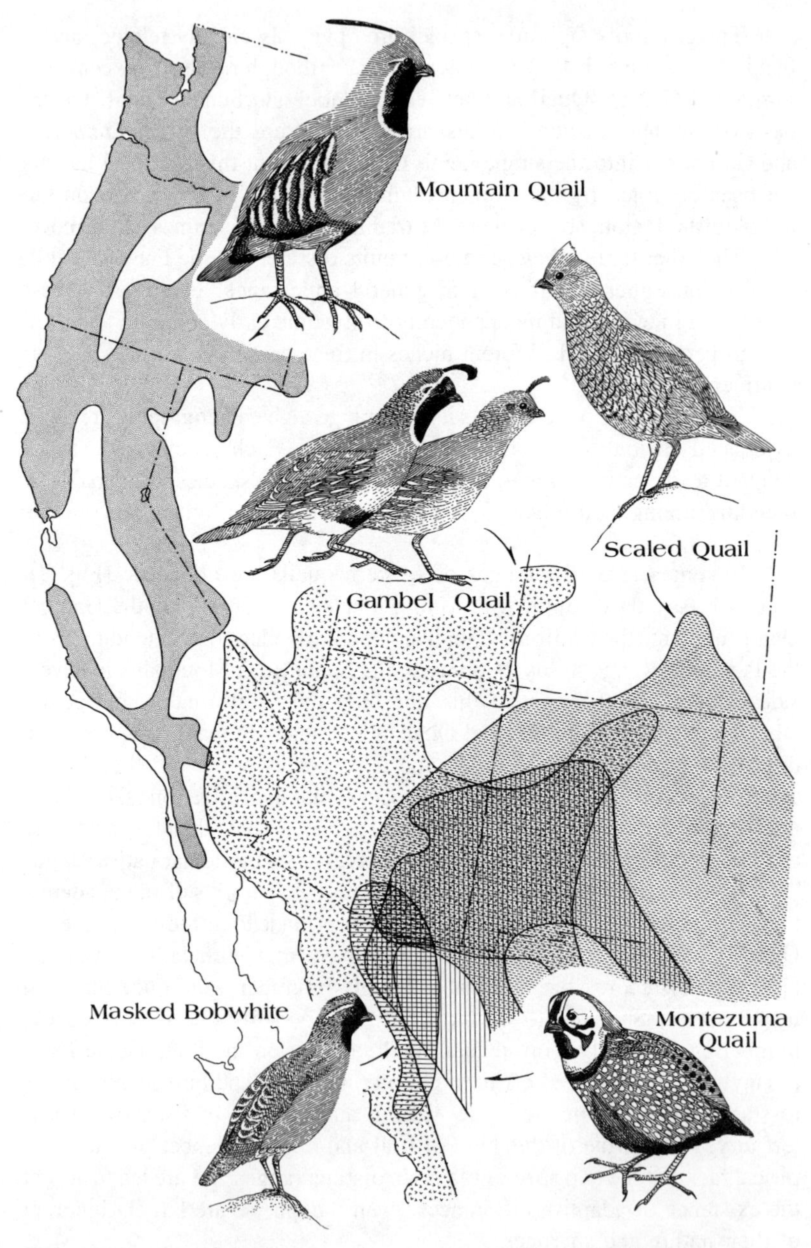

Figure 3. Ranges of five species of western quails. The California Quail range is shown in Fig. 2.

and 45°C Gambel Quail lost significantly higher percentages of water through evaporation than did Scaled Quail, meaning they had a more efficient cooling system. Both species were able to tolerate an ambient temperature of 40°C without ill effects, but at 45°C Gambel Quail survived better than Scaled Quail. Oxygen consumption values were similar at 30°C, but at 40° the values were significantly higher for Gambel Quail. Henderson (p. 436) concludes: "These results indicate that Gambel Quail are better adapted, physiologically, to hot arid environments than are Scaled Quail. This is quite plausible since Gambel Quail inhabit extreme desert areas while Scaled Quail are confined to more mesic areas with more moderate temperatures."

Several species whose ranges overlap (see Fig. 3) are separated on ecologic rather than geographic grounds. That is to say, they occupy different habitats in the same general area. This situation is well illustrated by the relationship between the California Quail and the Mountain Quail. The ranges of these two species overlap substantially. But on the ground there is actually a fairly clear differentiation of habitat with much less overlap than the map would indicate. The Mountain Quail occurs largely in conifer or oak timber, or in dense chaparral. The California Quail utilizes these cover types but usually frequents openings or edge situations rather than continuous dense cover. In some areas, Mountain Quail make substantial vertical migrations, moving well up into mountain forests in summer, dropping down to lower snowless zones in winter. On the west slope of the Sierra Nevada, for example, these seasonal movements may extend over distances of 50 miles or more. In winter the two species often are associated on the same ranges, and at times may even occur in mixed flocks. Yet it is clear that the Mountain and California quails are differently adapted and have evolved to utilize different habitats in close proximity to one another with a minimum of overlap or competition.

By the same token, the Montezuma Quail, Scaled Quail, Gambel Quail and the Masked Bobwhite are ecologically separated, although their gross ranges overlap.

SUBSPECIES OF CALIFORNIA QUAIL

The ancestral range of the California Quail, extending nearly 1300 miles from north to south and 300 miles from west to east, embraces some very distinctive eco-types. It is a reasonable presumption that segments of the total California Quail population are in turn differentially adapted to local habitats. Even though all individuals are members of the one species and there is no known barrier to their free interbreeding, still the local populations are not identical in appearance or in habit. Taxonomists label recognizable sub-populations of a species as geographic races or subspecies, and designate them by a trinomial Latin name. Thus the Valley California Quail, *Lophortyx californicus californicus,* occupies most of central California,

TABLE 1.
Average weights of California Quail from various parts of the native range.*

Subspecies	*Locality*	*Nearest parallel of latitude*	*Average weight (grams)*	*Number weighed*
L. c. brunnescens	Contra Costa Co., Calif.	38°	187.8	321
L. c. brunnescens	San Mateo Co., Calif.	37°	189.5	652
L. c. californicus	Yolo Co., Calif.	38°	175.6	64
L. c. californicus	San Luis Obispo Co., Calif.	36°	177.0	227
L. c. californicus	Los Angeles Co., Calif.	34°	157.3	29
L. c. plumbeus	Baja California	31°	157.7	57
L. c. achrusterus	Southern Baja California	28°	150.6	32

*Data derived in part from Sumner (1935:249)

whereas the Coast California Quail, *L.c. brunnescens,* is the form occurring along the humid coastal strip. Differentiation of subspecies has traditionally been based on morphologic characters that are observably different in preserved museum specimens. Birds of the race *brunnescens,* for example, are larger and demonstrably darker and more richly colored than the pale birds of the race *californicus*. The assumption is implicit that there are many other subtle genetic differences between subspecies that adapt them to their local habitats, and the morphologic differences are merely "markers" for the physiologic and behavioral adaptations that may be functionally far more important. Nevertheless, it is on the basis of the "markers" that most or all subspecies of birds have been described. Table 1 illustrates a gradation in size, the larger birds occurring in the north and along the coast, the smallest in the south and inland, following "Bergmann's Rule."

There has been considerable criticism of the fad to attach trinomial names to scarcely recognizable sub-populations of birds. Mayr (1951:94) observed that, "Instead of expending their energy on the describing and naming of trifling subspecies, bird taxonomists might well devote more attention to the evaluation of trends in variation." Nevertheless, the concept that sub-populations do differ genetically is a valid one, and whether we recognize this fact through use of trinomials or through some more descriptive device is not crucial to our consideration of the California Quail. I will use the recognized designation of subspecies, as authenticated by the Check-list of North American Birds, published by the American Ornithologists' Union (1957).

There follows a brief synopsis of the seven recognized subspecies of *Lophortyx californicus* whose ranges are depicted in Figure 4.

1. *L.c. californicus*. Valley California Quail. The most widespread race, its range extending from San Diego County, California, north into southern Oregon (excepting only the coastal strip from San Mateo

County to the Oregon border), the Central Valley, and adjoining slopes of the coast ranges and Sierra Nevada.

2. *L. c. brunnescens*. Coast California Quail. West slope of the coast ranges from San Mateo County to Del Norte County.
3. *L. c. canfieldae*. Inyo California Quail. East slope of the southern Sierra Nevada in Inyo and Mono Counties, California, and the adjoining fringe of west-central Nevada.
4. *L. c. catalinensis*. Catalina Island California Quail. Native only on Santa Catalina Island, Los Angeles County. Introduced on Santa Rosa and Santa Cruz Islands.
5. *L. c. plumbeus*. San Quintín California Quail. Southern San Diego County south into northern Baja California to approximately latitude 30°N (El Rosario).
6. *L. c. decoloratus*. Grinnell California Quail. Baja California, from latitude 30°N (El Rosario) south to latitude 25°N (El Refugio).
7. *L. c. achrusterus*. Cape California Quail. Southern tip of Baja California, from latitude 25° to Cabo San Lucas.

The characters which are used to differentiate these seven subspecies relate to slight differences in coloration and size which I will not attempt to describe here. Suffice it to say that accompanying the differences visible on preserved specimens may be some very significant behavioral and physiological differences that do not show at all. For example, a factor of undoubted importance in adapting a quail population to its home environment is the matter of timing of the breeding season. As will be brought out in another chapter, breeding behavior is initiated in quail, as in most birds, by seasonal changes in day length. This celestial trigger prepares the birds to produce chicks at the time of year most likely to be favorable for their survival. But the proper time for chicks to hatch may be very different in the deserts of Baja California than in the cool pine-sage uplands of Modoc County, or in the wet redwood-chaparral of Del Norte County. We presume, therefore, that each of these local populations has its own adjusted response to the lengthening days of spring; otherwise they would not have succeeded in persisting year in and year out through geologic ages.

MIXING GENETIC STRAINS THROUGH RESTOCKING

The concept of local genetic adaptation assumes considerable management significance when one considers the extent to which non-native breeding stocks have been liberated to augment dwindling wild populations. The peak of quail restocking in southern California occurred in the period 1933 to 1941. Liberations were obtained from three principal sources:

1. Birds reared on game farms.
2. Wild-trapped quail imported from Baja California.
3. Wild-trapped quail from refuge areas in southern California.

Presumably the original game farm stocks were derived from birds trapped somewhere in southern California, where the Los Serranos Game

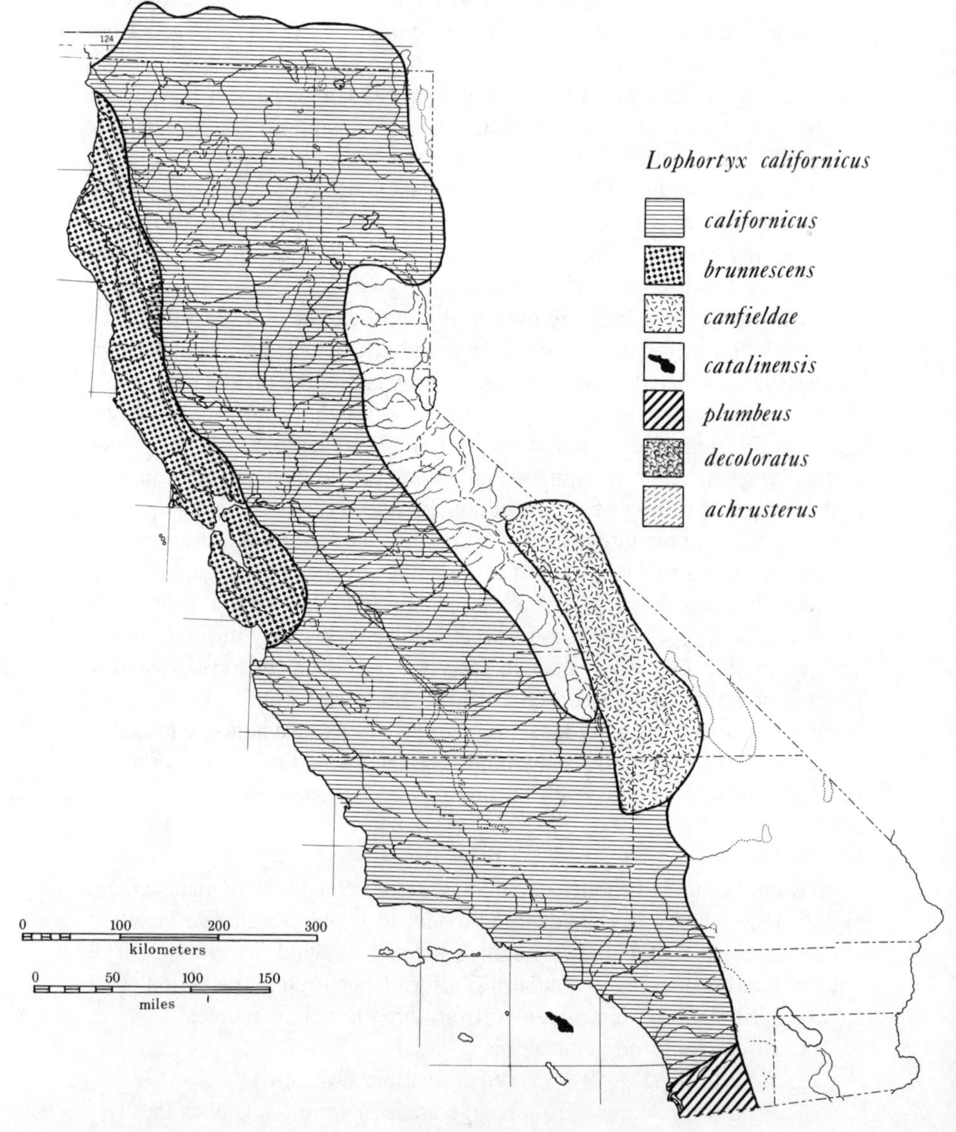

Figure 4. Original distribution of the subspecies of California Quail.

Farm was located. Likewise, wild-trapped native birds would be of more or less local genetic strains. The possibility of mixing non-adaptive genetic strains was greatest with the imported Mexican stocks.

True (1934) gives a fascinating account of the beginnings of the Mexican importation program (mentioned further in Chapter 3). In the autumn of 1933, the California Department of Fish and Game aproached the Mexican Department of Agriculture for permission to import wild-trapped California Quail. By December of that year, a permit had been arranged to export from Mexico 100,000 quail, and the first truckload of birds cleared the border on December 23, 1933. That winter, 8,297 live birds were imported, and, after a recovery period at Los Serranos Game Farm, 7,517 were released in the wild, while 780 were held as game farm breeding stock. The point of origin of this population was near the village of San Telmo, Baja California, about 200 miles below the international border. The subspecies of quail in that area is *Lophortyx californicus plumbeus,* the San Quintín California Quail. Here, then, we have evidence of a non-native genetic strain being liberated in southern California and additionally being propagated on the game farms for subsequent liberation.

What was the effect of this genetic mixing? Was the experiment in importation considered a success or failure? Unfortunately the history of this era in California's quail management program is obscure. Liberations from the game farm and from Mexican imports continued for a few years, but the number of birds liberated declined during the interval from 1936 to 1940 and apparently ceased some time thereafter. Richardson (1941) summarizes 791 band returns from some 65,000 quail banded and released in a nine-year period terminating in 1940. He does not draw any general con-

Figure 5. Typical quail habitat on Newhall Ranch, Ventura County. Cover grows on the background slopes, food on the more level terrace in the foreground. P. McBean photo.

clusions about the success of the quail restocking program, but it is significant perhaps that the endeavor was soon abandoned.

Whatever genetic maladaptations may have arisen from release of Mexican quail in California would have been quickly eliminated by natural selection favoring the adapted strain of the native stock. Presumably the Mexican quail crossed freely with the local birds, producing after a few generations an assortment of hybrids grading in characters from one parent stock to the other. Such a heterozygous population, exposed to the selective forces of the California environment, would in time surely revert toward the characteristics of the native race. Any ill effects of genetic mixings, therefore, can be viewed as temporary.

EXTENSION OF THE CALIFORNIA QUAIL RANGE THROUGH TRANSPLANTS

The California Quail has been successfully introduced to many areas where it was not native. Most notable is the extension of range in the Great Basin area, east and north of the ancestral range. As can be seen in Figure 2, the species now lives successfully in a very extensive area in Washington, Oregon, western Nevada, western Idaho, and in scattered localities in British Columbia, Utah, and eastern Nevada.

Transplants of California Quail to sites in the Great Basin were initiated at a surprisingly early date. The history in Nevada is elucidated by two newspaper accounts called to my attention by Bill Rollins of the Nevada Department of Fish and Game. The first of these appeared in the Carson City *Daily Appeal* on June 24, 1865:

QUAIL.—We were out in the hills about five miles from here, yesterday, and while riding over a wood-cutter's road in a little ravine that divides the bluffs west of the Mound house, scared up a hen quail and a dozen or more young ones, quite big enough to scamper away and hide in the sagebrush. These are the first quail we have known of in these parts. It is to be hoped that these birds may be suffered to escape the sportsman's gun. Indeed, they will escape from any *sportsman,* for all who are worthy of that title know and respect that section of the game laws of this State which protects quail and provides a penalty for persons found guilty of killing them. Where this solitary bird came from is a matter of conjecture. We believe it is decided by those who ought to know, that the quail is not indigenous to this part of the State. It is probable that the one we saw yesterday is an estray from somebody's cage. We are afraid that Madame Quail will find this a hard country to raise a family in.

An explanation for this occurrence is found in a second news release printed in the *Gold Hill Daily News* of Virginia City, May 4, 1878:

California Quail are becoming plentiful in several valleys in the western part of Nevada. The first of these birds were introduced into this portion of the country in 1862. On the 17th day of May of that year Lance Nightingale and Robert E. Lowery turned loose 22 California quail at Sol Giller's ranch, 4 miles east of Huffaker's

Figure 6. California Quail habitat in Lander County, Nevada. Willow, rose, big sage, and rabbit brush are characteristic cover plants. D. Erickson photo.

station. About 4 months later the same gentlemen liberated 18 more of these birds near Dutch Nick's ranch on the Carson River—just about where the town of Empire now stands. These birds were sent from California upon the order of Messrs. Nightingale and Lowery, and were brought up from Reno by Mr. Lowery. The *News* is under obligation to Mr. Lowery for the facts embodied in this item. Since Mr. Nightingale has "gone to the happy hunting grounds" it is only right that our sportsmen should know to whom they are in a great measure indebted for the plentiful supply of the game-some bird above mentioned.

The bluffs west of Mound House, mentioned in the first article, are about 2 miles northeast of Dutch Nick's ranch referred to in the second article. It is apparent that the release of quail made in 1862 was not widely publicized. The precise origin of the transplanted stock is not stated, but the probable source was central California, within the range of *L. c. californicus*. Linsdale (1936) identifies the present quail population of west-central Nevada as belonging to this race (known then as *vallicola*).

There is an additional record in 1910 of 60 birds transported from the Sacramento Valley by wagon and released in Paradise Valley in the extreme northwest corner of Nevada (William Molini, personal communication). According to Nissen (see Appendix A), there was no indigenous quail population in Paradise Valley. Thirty-seven additional transplants are recorded between 1955 and 1971, all made by the Nevada Department of Fish and Game. In almost all cases the released birds were wild-trapped in western Nevada, many within the city limits of Reno, derived presumably from the stock imported originally by Messrs. Nightingale and Lowery. A number of these transplants have been successful, notably along river valleys where appropriate habitat is available—riparian willow and rose thickets offering the best cover, and ranch feed-lots supplying crucial winter food. Today the California Quail is a common game bird in many localities in western and northern Nevada.

Of the introductions in British Columbia, James Hatter (pers. comm.) supplied the following information: "They [California Quail] were introduced to Vancouver Island and the Fraser Valley sometime during the 1870's and to the southern interior of British Columbia about 1900, according to Taverner (1934). Munro and Cowan (1947) maintain that the southern Okanagan was populated by birds moving northward from the State of Washington, where the original stock was also introduced." The source of the "original stock" is not given. Lewin (1965) says the first officially recorded introduction of quail in the Okanagan Valley occurred in 1912, but there is a record of a bird seen in 1911, indicating a previous release. By 1915 a short hunting season was declared in this area.

Dr. Hatter's report continues:

> Introductions to the Fraser Valley were not successful, although small numbers still maintain themselves there. In the southern interior they are abundant in certain areas: we found them plentiful around Osoyoos in the spring of 1950. On Vancouver Island the populations built up tremendously, and for a number of years the California quail was most abundant from Sooke north through favourable habitat to Comox. However, a population decline set in, and to-day the birds are maintained at a comparatively low level, although they are still locally abundant on parts of the Saanich Peninsula and in the City of Victoria. Many introductions follow a pattern of initial abundance followed by decline and a continued low population level, but the reasons for this phenomenon are not clearly understood.

We do not have details of the introduction of California Quail in Idaho. But of more recent history, Dick Norell (pers. comm.) writes:

> Valley quail have done exceptionally well in Idaho since the early 1960's and have occupied much of the former range of mountain quail which have been at low numbers for the past several years. There has been some speculation that valley quail have displaced mountain quail; however, I am of the opinion that valley quail have simply occupied the void created by the low cycle in the mountain quail population. There has been increased interest in quail hunting which coincides with the valley quail expansion. . . . [T]he quail harvest is averaging slightly over 100,000 since 1964. This is almost double the 1954–63 average of 53,000.

Concerning the introduction of California Quail in Washington, Fred Martinsen (pers. comm.) sends the following information: "These birds were brought in from California and were first released on the west side of the state in 1857. In 1914, the first releases occurred on the east side of the mountains. . . . This species is [now] well distributed throughout the state and frequents the brushy areas along creek and river bottoms. . . . Counties having the highest concentrations of these birds are Asotin, Benton, Chelan, Columbia, Okanogan and Yakima."

Anthony (1970a:276) adds this comment on the history of the California Quail in eastern Washington: "According to the records of the Washington Department of Game, the first introduction of 108 birds was made in Walla Walla County in 1914. . . . The California Quail presently attains relatively dense populations in canyons where shrubby vegetation still persists. Today it is one of the important game birds of this region."

Transfer of breeding stocks within the state of Oregon is thus described by Masson and Mace (1970:11):

The Valley quail is a native bird originally confined to the counties bordering California and Nevada. Two distinct subspecies were recognized, the California quail *(L.c. californica)* in Josephine and Jackson Counties and the Great Basin quail *(L. c. orecta)* in southern Klamath, Lake, and Harney Counties. Trapping and transplanting operations have extended their range to include most of eastern Oregon as well as many of the western Oregon counties. Today the two subspecies are mixed and no longer distinct. Highest densities of valley quail at present are found in the Columbia Basin and in central and southeastern Oregon.

The race *orecta* is no longer recognized as valid, so all Oregon birds probably derived from the nominate race, *L. c. californicus*.

Rawley and Bailey (1972:21) comment thus on the species in Utah:

The California quail is native to states of the Pacific Coast. It was first introduced into Utah in 1869. Subsequent releases, trapping and transplanting, and dispersion have resulted in establishment in many parts of the State. Heavy snows limit their food supply. They build up in numbers through a period of mild winters only to be depleted in severe winters. It is hunted as an upland game bird, producing excellent gunning in some local areas.

David E. Brown of the Arizona Game and Fish Department tells me (pers. comm.) that there is an incipient population of California Quail becoming established near the Lyman Reservoir on the Little Colorado River in Apache County, Arizona. This population resulted from a private introduction of stock obtained from Oregon in 1960, and shows promise of persisting and spreading.

It is clear from the successful introduction of the California Quail through the northern Great Basin and elsewhere in the west that the species has the genetic capacity to adapt to new range. What, then, limited its spread into these regions originally? It would appear that the clearing of forests and the introduction of grain crops, annual weeds, livestock feed lots, and irrigation created favorable habitat that did not exist previously. Of the Okanagan population, Lewin (1965:62) states: "Quail are definitely associated with orchards and irrigated areas and are rarely found above the 2000-foot contour line. Generally, irrigation is not practiced above this elevation and the habitat which is of the ponderosa pine-grassland type is not suitable to quail. The valley in its original primitive condition would have provided little quail habitat." This is an interesting point, since in California—the

Figure 7. Coastal scrub near Bodega Bay is improved for quail occupancy by cattle grazing.

original heart of the species range—land use practices stimulated an initial increase in quail but this was not sustained. The history of the rise and fall of quail abundance in California will be recounted in the next chapter.

Another possible factor limiting the northward and eastward spread of the quail would be the existence of geographic or ecologic barriers, such as mountains or arid plateaus, that precluded successful emigration. Transplanting breeding stocks is a device for circumventing barriers to natural spread.

THE CALIFORNIA QUAIL THROUGHOUT THE WORLD

In addition to extending the range of the California Quail in North America, attempts have been made to introduce the species in many other parts of the world, some of which have succeeded.

The quail is now established on various islands in the Hawaiian chain, stemming from an original transplant in 1855. Schwartz and Schwartz (1949) give an excellent account of the status of the quail as of the date of writing. Major populations occurred on the drier (leeward) portions of Hawaii and Molokai, with lesser numbers on three of the other islands. Curiously, the stock on Hawaii was derived from the Valley California Quail *(L. c. californicus),* while that on Molokai came from the Coastal California Quail *(L. c. brunnescens)*. On other islands the two races appear to have been mixed. The Schwartzes estimated a population of 77,000 birds on 1,800 square miles of occupied range, for an average density of 43 birds to the square mile, which is a low population by California standards. Some local densities reached 600/sq. mi., or a bird per acre, which is comparable to thrifty populations in California.

In New Zealand, the species has been moderately successful. Williams (1952) gives a detailed account of the introduction and subsequent establishment on both the North and South Islands. Initial stocking occurred in 1862 near Auckland. From this point, birds were trapped and moved to many parts of New Zealand, largely by the "acclimatization societies" that were responsible for bringing many exotic birds and mammals to the islands. The species now occurs locally throughout much of North Island, and on South Island in Central Otago and in scattered localities north and east of the Alps. A series of 29 study skins taken from near Lake Taupo and in Central Otago have been identified as *L. c. californicus;* a single skin (loc. unknown) was previously identified as *L. c. brunnescens*. Both races may have been introduced. In one relatively arid area in Central Otago, Williams reports local densities varying seasonally from 1.2 to 3.4 birds per acre. This is the only area in New Zealand where the species could be considered abundant.

New Zealand birds have been shipped to other South Pacific islands (Fiji, Tonga, Norfolk, Chatham Islands) and to various points in Australia. Williams states that small numbers are reported to persist on Chatham Islands. Harry Frith (pers. comm.) reports a thriving population on King Island in Bass Strait, but other introductions in Australia apparently failed.

Another successful transplant occurred in central South America. De Shauensee (1966) reports that the species is well established in western Argentina (San Juan and Mendoza) and central Chile (Coquimbo to Concepción). A population also exists on Mas Atierra Island of the Juan Fernandez group off the west coast of Chile. No details are available of the race or races of quail involved, nor of population densities.

In summary, the California Quail has demonstrated considerable adaptability to a wide variety of habitats in the temperate and sub-tropical regions of the world, made suitable by patterns of land development and human use. This suggests that the highly local adaptations to climate and habitat, alluded to earlier in this chapter, are not so fixed as to preclude adjustment to other favorable environments.

COMPETITION WITH EXOTIC GAME BIRDS

One management program that may affect the status of the California Quail is the introduction of exotic gallinaceous birds. The fear is frequently expressed that exotics may interject themselves into the niche occupied by the native quail and force a decrease in the latter. What evidence is there to support or reject this inference?

The two established exotics that now share range with the California Quail are the Ring-necked Pheasant and the Chukar Partridge. Both have become firmly ensconced as California game birds, and both are locally numerous in some areas that were once the specific domain of the quail.

Pheasants were liberated in California from 1889 on, but only became well established with the development of irrigated grain crops in the Central Valley in the 1920's. Rice especially seemed to create favorable pheasant habitat, and the development of large-scale rice agriculture in the Sacramento Valley following World War I stimulated a virtual explosion of the pheasant population. Glading (1946:168) states: "This change in agriculture caused a marked change in habitat conditions as far as pheasants were concerned. Our present huntable crop of pheasants in California is largely tied up with ricelands and other grainlands in irrigated areas. . . ."

The rich valley lands occupied by pheasants had originally been the domain of the California Quail. Riparian woods and thickets adjoining cultivated grainlands were at one time heavily populated with quail, giving rise to the vernacular name Valley Quail. However, the growing intensity of agriculture led to the uprooting of wild woody vegetation which is essential for quail cover. Even before the pheasants became well established, quail were steadily decreasing in the cleared valley lands. When ultimately the pheasant had become the predominant game bird of the valleys, and quail had become scarce, it was impossible to say that the pheasant had in any way been instrumental in eliminating the quail. Rather, it was much more plausible to believe that a major change in the landscape destroyed the quail habitat while concurrently creating ideal pheasant habitat. This interpretation, stated by Glading in 1946 (p. 175), concurs with my thinking. Even where quail and pheasants still occur together in marginal range situations, I see no evidence of direct competition. These two species are too different in habitat needs to be serious competitors.

Ecologic differentiation is not so clear in the case of the Chukar Partridge. The Department of Fish and Game began liberating chukars in 1932. By 1950 the species was well established in the semi-arid range lands of the upper San Joaquin Valley, and bordering the Sierra Nevada on the north, east, and south. (See Harper et al., 1958:8 for range map.) At many points in its range, the chukar now shares habitat with the California Quail and elsewhere with the Gambel Quail. Chukars and quail often drink at the same water holes. There is a potential for direct competition between the exotic partridge and the native quail.

The subject is discussed by Harper et al. (1958:34) in their comprehensive report on "The Chukar Partridge in California." I quote this portion of their report in full:

> Evidence of chukars competing or conflicting with native game to any serious extent was not indicated during the study. The only competition which might occur would be in areas where water supplies are limited, such as at guzzlers, where the problem could be easily eliminated by placing double units or larger containers in the area. Valley quail, mountain quail, doves, rabbits, and other wild life were observed to use the same water source compatibly. Only one instance of conflct was recorded—an adult chukar was observed killing a Gambel quail chick at a waterhole.

Occasionally, groups of chukars with a single quail in their midst were seen, and in several instances a single chukar in a covey of valley quail was observed. Such instances are thought to result from the hatching of quail or chukar eggs in the nests of the other species.

Competition for food was not prevalent, since the chukar utilizes the more open knolls and flats, which are not frequented by other game.

As in the case of the pheasant, I concur with the opinion that competition between the introduced chukar and the native quail is not critical. The two species seek quite different types of cover. They overlap some in food habits, particularly in semi-arid areas such as the Temblor Mountains, where filaree *(Erodium)* and fiddle-neck *(Amsinckia)* are staple items of diet for both chukar and quail. There is less overlap in desert situations, where Harper et al. (1958) found Russian thistle *(Salsola)* to be a key chukar food.

In years of good rainfall, when breeding conditions are optimum, both chukars and quail can ''explode'' to high population levels without apparent mutual interference. In dry years, both species seem to persist in small numbers, or if one species drops out it is the chukar, not the quail.

It would appear that the introduction of exotic gallinaceous birds has not created significant competition for the California Quail.

2
HISTORICAL PERSPECTIVES

INDIAN USE OF CALIFORNIA QUAIL

A number of aboriginal tribes in California trapped and hunted quail, both for food and for feathers used in decorating baskets and ceremonial costumes. The skills involved in capturing quail apparently were best developed among the Miwok, Pomo, Maidu, and Patwin tribes of central California, from the Sierra foothills to the coastal ranges around Clear Lake. However, there were quail hunters as well among groups living far to the southward, and it seems safe to say that, among native terrestrial birds, quail were the most extensively used. Waterfowl were hunted even more heavily, where they occurred in abundance.

Karen M. Nissen has assembled a comprehensive summary of the utilization of California Quail by aborigines, and her report, entitled "Quail in Aboriginal California," appears as an original research publication in Appendix A of this volume. The surprisingly rich literature on the place of quail in the culture of California Indians yields some suggestive information on quail abundance as well as quail use.

THE QUAIL IN COLONIAL CALIFORNIA

The first Spanish settlement in California was established in 1769 in the vicinity of San Diego, and other missions to the northward followed in rapid order. However, the good friars paid scant attention to small birds, and their chronicles of the California missions tell us little or nothing about the quail. In the vicinity of each mission, European livestock was grazed and lands were broken for cultivation. With the introduction of crops came many Mediterranean annual weeds. The quail habitat was certainly modified and probably much improved by Spanish agriculture close to the missions, but of this we can only surmise.

First mention of the California Quail is found in the chronicles of Jean Francois Galaup de la Pérouse, who visited Monterey in September of 1786. In the three-volume account of his voyage around the world, he includes a remarkably accurate plate of the California Quail (Fig. 8) and comments on its occurrence and use by the native Indians: "The brush country and the plains are covered with little crested gray partridges which, like those of Europe, live gregariously, but in coveys of three or four hundred. They are fat and of good flavor" (Rudkin trans., 1959:52). Of native hunting, he states: "The natives of Monterey, small, thin, and colored somewhat like Negroes, are very adept archers. The French saw them hit the smallest birds. These they stalk with the most amazing patience; . . . shooting only when they are about fifteen paces away" (Gassner trans., 1969:46). There is no evidence that specimens of the quail were transported to France by La Pérouse, but one wonders about the accuracy of the plate, which presumably was rendered upon return from the voyage.

A few years after the French stopover in Monterey, the area was visited by a British expedition. Archibald Menzies was the first trained naturalist to observe the California Quail. In 1790, he accompanied Captain Vancouver on his voyage around the world in the ship "Discovery," serving in the capacity of Naturalist and Surgeon. During stops along the Pacific

Figure 8. The first illustration of the California Quail, appearing in the published report of J. F. G. de la Pérouse (1798) concerning his journey to the New World.

coast, Menzies went ashore whenever circumstances permitted, and his full and meticulous notes, still preserved in the British Museum, contain many original observations of the fauna and flora of this new land. His first description of the California Quail was recorded at Monterey, on December 5, 1792, as follows:

> The Thickets every where were inhabited by great variety of the feathered Tribe, many of which were also new, among these was a species of Quail of a dark lead colour beautifully speckled with black white and ferrugeneous colours with a Crest of reverted black feathers on the crown of its head, these were also met with at Port San Francisco and are common over this Country, they are equal to the common Partridge (English gray partridge) in delicacy of flavour and afforded a pleasing variety to the other luxuries with which at this time our Table abounded. (Grinnell, 1932:246)

Menzies probably collected the type specimen of California Quail during that sojourn in Monterey. The species subsequently was described by George Shaw and Francis P. Nodder in Volume 9 of the *Naturalists' Miscellany,* 1797.

As previously noted by La Pérouse, the native Indians of the Monterey area made some use of the local quail, for Menzies wrote: "Their food at this time was chiefly shell fish, which the Women collected along shore, while the Men lounged about the Country with their Bows and Arrows, killing Rabbets and Quails, which they generally brought to us to barter for beads and other trinkets" (ibid., p. 247).

At Santa Barbara he noted: "The thickets swarmed with squirrels and quails and a variety of other birds which afforded some amusement in shooting them as I went along" (ibid., p. 249).

Again at San Diego he commented on quail: "The following day I accompanied a party of the Officers to the Peninsula between the Harbour and the Bay, where we traversd over an extensive plain beating up the thickets in quest of game and though a number of hares rabbets and quails were seen, yet being in the heat of the day, they lurkd so close in amongst the brush wood that we had but very indifferent success . . ." (ibid., p. 251).

The trappers and mountain men that subsequently came to California were interested only in furs and in the larger game animals. They lived handsomely on the native deer and elk of the state and occasionally on domestic livestock that over-ran the rich and productive rangelands. Significantly, it was after the gold rush had subsided, leaving the country fairly destitute of edible larger mammals, that the settlers and homesteaders began to make use of smaller game for food. Only then do we find written reference to the quail as a game bird and as a food resource.

MARKET HUNTING DURING THE PERIOD OF SETTLEMENT

The most valuable chronicles of the market hunting era are left to us by Mr. Walter R. Welch, who in the latter years of his life published several articles in *California Fish and Game* on his youthful experiences as a market gunner. We can do no better than to quote Mr. Welch directly and at length (1931a:255):

> Although I was born at Olema, Marin County, my first recollection of wild life begins in the Sunol Valley, Alameda County, in 1866. At that time quail in abundance were to be seen throughout the whole valley. . . . In 1867 we moved to a ranch located between "Spanishtown" now called Half Moon Bay, and San Gregorio, on the coast side of San Mateo County. There I saw quail by the thousands everywhere; every canyon, gulch and ravine contained quail . . . and the whole country seemed to be alive with them. . . . The farmers' wives and children, and other people trapped the quail, and market hunters shot them and shipped them to San Francisco markets, where they found ready sale—with the result that from the early sixties to the late eighties quail traps by the hundreds could be seen throughout the entire country. In 1872 we moved to a fruit orchard in the vicinity of Mayfield, Santa Clara county.
>
> At that time, where Stanford University and Palo Alto are now located, were unbroken fields covered with live oaks, chaparral, manzanita, and Toyon berry bushes, alive with quail and other small game. . . .

In another article devoted specifically to the history of commercial quail hunting in central California, Welch continues (1928:123):

> What a difference in the supply of quail in 1872 from that of 1927. The "Butchart boys" began the hunting of quail for the San Francisco markets in Marin County during the sixties, and about 1870 moved to San Mateo County, locating at San Gregorio.
>
> The Butcharts were brothers, and were real sportsmen of the old Scotch school. While they marketed the game they killed, they did all their shooting on the wing, over well-trained bird dogs. They would spend more time to retrieve a wounded or dead bird than would be required to flush and kill a half dozen birds. Both were splendid wing shots, frequently bagging as many as 60 quail each in a day's shoot. I have seen Jim Butchart kill 27 quail straight, singles and doubles, without losing a bird, the birds being flushed from brush cover.
>
> The "Bissell boys," Englishmen, and twin brothers, were also early-day market hunters for quail in San Mateo County. They did their hunting on the coast side of the county, and also in the vicinity of Woodside, west of Redwood City. The Bissell boys were also good wing shots and did their shooting over dogs. . . .
>
> While those mentioned were the recognized market hunters for quail in San Mateo County during the early days, there were many others who shot and trapped quail for market in that section of the state. Some of those who trapped quail maintained a string, often consisting of as many as fifty traps.
>
> The figure four set was the trap commonly used for trapping quail, much of the trapping being done by boys, and the wives of farmers and ranchers. During the sixties and seventies, and even as late as the early eighties, traps set for quail could be seen all over San Mateo County. All that was required to construct a trap capable of catching quail was a strong, sharp pocket knife with which to cut hazel or other material about one-inch in diameter by four feet in length, which would be placed one upon another in pyramid form, the pieces being made shorter as the trap was built up. When the trap was completed it would be held together by a "binder"

across the top, which was fastened by a piece of string to the bottom slat of the trap, thus binding the trap firmly together; the trap being baited with wheat or other grain or seed that might be obtained.

During the sixties, seventies, and eighties, Dr. R. O. Tripp, who conducted a general merchandise store at Woodside, San Mateo County, and who made regular weekly trips by team and train between Woodside and San Francisco, handled nearly all of the quail killed by market hunters in that section of the country. Dr. Tripp paid the hunters cash for their quail and resold the birds to the retail game dealers in San Francisco. . . .

As the means of transportation developed, the market hunting and trapping of quail spread down the peninsula into Monterey and other southern counties, where, during the late eighties and until about 1901, in the section of country west of Bradley and King City, and in the vicinity of Jolon, Pleyto, Poso, San Ardo, Paso Robles and Santa Margarita, thousands of quail were slaughtered each year by market hunters. In these sections of the state great numbers of quail were destroyed by nets used to trap quail at springs and water holes. The trappers would use a piece of 1-inch mesh net about 15 by 50 feet. The back part of the net was usually buried in dirt, the ends and sides of the net were fastened to willow poles, and a stick about 2 feet in length, to which was fastened about 200 feet of strong string, was used to hold up and trip the trap. A pan filled with water was placed under the net and the trap baited with wheat, chaff or other grain or seed.

As water was scarce in the section of country mentioned, quail in bands consisting of thousands of birds would congregate about the springs and water holes in the vicinity of which the traps were set. When a sufficient number of quail had entered the net, the trapper who was watching the trap would pull the stick from under the net, and thus cause the net to drop on the birds.

In this manner at times quail to the number of several hundred would be caught by one fall of the net or trap. The trapper would then proceed to kill the birds by using a willow stick with which to hit the birds on the head as they stuck out through the meshes of the net. After the birds had been killed and removed from the net, the net would be reset and baited.

The trapper would then proceed to draw the quail and tie them in bunches of six birds each. The bunches of birds would be hung up in a tree at a distance of about 30 yards, and a shot from a shotgun fired at them, in order to remove evidence that the birds had been trapped.

As a rule the trappers would fill the springs and water holes with brush and so block it that quail would not be able to secure a drink. The net would then be set, baited and so left for a day in order that the quail would become accustomed to it and enter it; then the trap would be sprung—usually twice in one day—which would result in about cleaning up the flock of quail that frequented that particular locality, which might consist of several hundred birds. The trapper would then move to another location. In this way he would be able to trap and ship several hundred quail twice per week.

As a rule the quail trappers above described worked in pairs, and their outfit usually consisted of a pack horse or mule, a roll of blankets, coffee pot, frying pan, a short-handled shovel, an ax, a shotgun, a piece of 1-inch mesh 15 by 50 feet, a quantity of wheat or other grain for bait, and a large, shallow pan to be filled with water and placed under the net or trap. . . .

The price paid for quail ranged from $0.50 to $1.75 per dozen depending upon

Figure 9. Poster depicting market hunting for quail, circa 1900. California Academy of Sciences print.

the condition of the birds. In 1885, I saw many quail exposed and offered for sale at grocery stores and butcher shops on Third Street, San Francisco, for 50 cents per dozen.

The quail trappers in the San Joaquin Valley, and also in the Monterey County section of the state, seldom did any wing shooting at quail. When quail were shot at in these sections, the shooting was done by "ground sluicing" the birds at springs and water holes, when from 40 to 60 quail would be killed at one shot.

It was not until 1895 that a decrease in the supply of quail became noticeable, and it became apparent that something must be done, and done quickly, to protect quail or the supply would be totally destroyed. Therefore, the Fish and Game

Commission, in order to be supplied with data to support the passage of needed legislation, detailed a man to ascertain the number of quail sold in the Los Angeles and San Francisco markets in 1895–1896. The figures secured showed that during the open quail-shooting season for 1895–1896, 177,366 quail were sold in the open markets of Los Angeles and San Francisco alone. Of this great number of birds Monterey County furnished 39,831; San Luis Obispo, 25,526; San Bernardino, 12,663; and Los Angeles, 11,026. From various other counties in the state were shipped to and sold in the two cities, quail ranging in number from 89 to 9800. These figures do not include quail shipped to and sold in the various other cities and towns within the state, nor quail killed and consumed by the hunters, their families and their friends. The 177,366 quail sold for a total of $15,160.08, or at an average of less than 10 cents apiece.

These figures certainly furnish the present generation and those who are inclined to lay the cause for the present scarcity of quail at the door of predatory birds and animals, with food for thought.

It is quite safe to say that during the eighties and nineties not thousands, but millions of quail were shot, trapped and sold in California, and that had it not been for the enactment of a law in 1901, which fixed a bag limit and prohibited the sale of quail, the supply of quail in this state would have been totally exterminated. The law enacted in 1901 fixed the bag limit at 25 per day and prohibited the sale of quail.

Subsequent to the passage of this law the market hunters resorted to bootleg methods in shipping quail. During the years 1901 to 1909, in order to be able to secure a sufficient number of quail to warrant shipment in a trunk checked as baggage, hunters to the number of eight or ten in the Monterey section of the state would combine in a quail hunt, pool their bag of quail, and in that manner be able to escape game wardens in the field, and transport large numbers of quail to the large cities. Quail were shipped to San Francisco and other large cities in egg cases, butter boxes, rolls of blankets, green cow hides, suit cases; in boxed demijohns as wine, in coal oil cans as honey, in kegs as butter and salt fish, and in trunks checked or expressed as baggage. Here they sold at from $2.50 to $6 per dozen. In this way, for a number of years despite the activity of game wardens, thousands of quail found their way to the markets of the large cities and were sold in restaurants and hotels, with the result that by 1925 only a pitiful remnant remained of this state's once bountiful supply. . . .

It is not predatory birds and animals, or inbreeding, that is responsible for the decrease in the supply of quail in California. The scarcity of quail in this state at this time is due to the usurpation of their food and cover, and to man, who has thoughtlessly and indiscriminately killed and destroyed quail at all seasons of the year in violation of law.

Grinnell, Bryant, and Storer (1918:534) compiled a good deal of historical information on the quail, for use in their classic book, *The Game Birds of California*. Some excerpts from that work follow:

W. T. Martin of Pomona states that in 1881–84 he and a partner hunted Valley Quail in Los Angeles and San Bernardino counties for the San Francisco markets. Eight to fourteen dozen were secured daily, and in the fall of 1883 the two men secured 300 dozen in seventeen days. Martin himself secured 114 birds in one day's hunt. In 1881 and 1882 over 32,000 dozen quail were shipped to San Francisco from Los Angeles and San Bernardino counties, and brought to the hunters engaged in the business one dollar a dozen. In those days restaurants charged thirty cents for quail-on-toast. By 1885 hunting had become unprofitable because of the reduction in the numbers of quail.

A. E. Skelton, of El Portal, Mariposa County, tells us (MS) that years ago when he was hunting for the market in the vicinity of Raymond, Madera County, he averaged about sixty birds per day. By careful handling he was able to secure better prices than other market hunters. After killing a dozen or fifteen quail, they were drawn, tied three in a bunch, and hung up to cool over night. All the birds which he had thus prepared were on the following day placed in wooden boxes with thin boards between each two layers. They thus reached the cities in beautiful condition and brought him from $1.50 to $2.25 per dozen, fifty cents more per dozen than quail shipped loosely in sacks, as was the practice of other hunters.

T. S. Van Dyke (1890, p. 460) states that market hunters used to ship 10,000 quail apiece during a single season; daily bags of 200, made by sporting men shooting the birds singly on the wing, were not unusual. C. H. Shinn (1890, p. 464) says that in eighteen consecutive hunts two hunters at San Diego secured from 47 to 187 quail on each hunt, in addition to other game; six bags of more than one hundred each were made. Other individual daily bags of six, twelve, and twenty-two dozen, respectively were known to this author. In the hills between the southern San Joaquin Valley and Carrizo Plains, E. W. Nelson (A. K. Fisher, 1893a, pp. 28–29) found the Valley Quail very numerous.

It was excessively abundant at some of the springs in the hills about the Temploa Mountains and Carrizon Plain. In the week following the expiration of the closed season, two men, pot-hunting for the market, were reported to have killed 8,400 quail at a solitary spring in the Temploa [Temblor] Mountains. The men built a brush blind near the spring which was the only water within a distance of 20 miles, and as evening approached the quails came to it by thousands. One of Mr. Nelson's informants who saw the birds at this place stated that the ground all about the water was covered by a compact body of quails, so that the hunters mowed them down by the score at every discharge.

Not only were quail shot for the market, but previous to 1880, they were regularly trapped in large numbers. In that year the practice was stopped by law. Cooper, writing in 1870 (1870a, p. 551), states that they were constantly exposed for sale alive in San Francisco, where many escaped from their cages to fly from roof to roof, occasionally descending into city gardens. Many trapped birds were shipped east at that time.

Howard Twining interviewed seven market hunters who operated in Contra Costa County in the period 1880 to 1901. He summarized what he learned from these old-timers in an article published in *California Fish and Game* (Twining, 1939:30):

Quail shooting started in September. By this time the young were full grown and family groups had combined to make large coveys. Hunting was best in cold damp weather when coveys drew together to make large flocks of several hundred. Also, it was more comfortable to hike in cold weather and the dog worked more effectively. All the hunters I interviewed hunted exclusively with 12-gauge shotguns. They used brass shells which they loaded with 3¼ drams of black powder, two hair wads, one ounce of number 8 shot and a wad to seal the end. The cost was about one cent a shell. Mr. Bart Gerow ground up pellets of blasting powder in a coffee grinder, cut his own wadding, and got his shot from a bottle washing plant, which cut his expense to about one-quarter of a cent a shot. . . .

A good dog was of course indispensable, but certain methods were used to supplement the work of the dog. Flocks were located by the use of a "quail caller." This was a green stick about the size of a lead pencil with a split from one end to the middle. A green laurel leaf was inserted into the split and the edges of the leaf were trimmed flush with the surface of the stick. When held to the lips and blown, the "ca-ca-coo" call of the quail could be easily imitated and answers would be received from members of every flock within hearing distance. The hunter tried to get between birds in the open field and the brush. A quail that had "frozen" in the grass and would not fly could often be raised by the hunter if he would kick a bush and blow through his lips making a whirring sound like a flying quail. Any quail within fifteen feet would immediately fly to join the bird it supposed had flown.

As soon as there was a lull in the shooting, the quail were cleaned. One deft movement of the finger disemboweled a quail so a knife was not necessary. After the day's hunt the birds were strung on a string and hung in a cool place. Twenty-five to 30 birds was a good day's bag, with usually 6 to 12 cottontail or brush rabbits in addition. Quail brought $1 to $1.75 a dozen in Oakland markets. Rabbits brought $1.25 to $1.50 a dozen. Ground squirrels were not found in quail habitat, and usually so many were bagged that their weight and bulk made it difficult for the hunter to make his way through brushy country, so squirrel hunting was carried on separately. Usually about eight to ten dozen squirrels were shot in a day, so it was necessary to make caches at intervals to be picked up at the end of the day's hunt. Squirrels brought the hunters $0.60 a dozen unskinned or $1 a dozen skinned. They were sold by the markets as "rabbits." The hunters shot five days a week and took their game to market on Wednesdays and Saturdays. After market hunting for quail was prohibited in 1901, the birds brought poachers $5 to $6 a dozen. One hunter said that he had hunted for the market as late as 1922.

Hunting stopped in March when quail coveys broke up and the birds started to pair off. Ground squirrels at this time became too fat to make good food. Haying season arrived in April and labor then was in demand, so there was only a short time interval between periods of steady employment for the market hunters. . . .

Although estimates varied, all [hunters] agreed that quail numbers had decreased continually throughout their experience. All believed that the heavy drain on the quail population by market hunters contributed greatly to depletion, but most of them were at a loss to explain why quail have not come back since market hunting was outlawed in 1901, and especially in the last few years when farmers have allowed very little hunting by sportsmen in the Lafayette region. A large part of the hills near Lafayette is owned by a water company, and hunting there is strictly forbidden, yet quail seem to be no more numerous there than in other regions.

Mr. Bill Gerow claimed that he noticed a decided drop in quail numbers from 1880 to 1885, coincident with the introduction of breech-loading shotguns, and with the conducting of extensive squirrel poisoning campaigns.

Many additional notes on the era of market shooting could be added, but the above quotes present an adequate picture of the situation. From these accounts I distill two questions of major interest in our present endeavor to understand and manage the California Quail populations:

1. What fortuitous combination of ecologic forces led to the explosion of quail numbers in the era 1860 to 1895?

2. What factor or factors broke the chain of production, starting in the late 1890's, leading to the marked decrease of quail all over California?

If we can ascribe plausible answers to these questions, we shall be well on our way to defining the foundation of a management program.

THE QUAIL "PEAK"—1860 TO 1895

It seems highly unlikely that quail in the pre-settlement stage were as abundant as they were subsequently during the market hunting era just described. I envision a sequence of environmental stages, associated with settlement and agricultural development, that initially favored the increase and spread of quail but later led to habitat deterioration and a substantial regression in numbers. Perhaps the historical sequence was parallel to that described by A. Leopold (1931:24) for the Bobwhite Quail in the north central states:

> To visualize the history of quail in the north central region, the reader must picture four distinct stages. These did not occur in all parts of the region at the same times, but they occurred in the same order, and the status of quail changed greatly with each successive one.
>
> First comes the virgin or pre-settlement stage. Our knowledge of this is conjectural only. It seems likely that quail were confined mostly to the edges of prairies, and to open woods made park-like by frequent fires. There were probably severe fluctuations in abundance, and also distinct seasonal movements by reason of changes in weather, fire, mast, seed crops, and predators.
>
> Next came the era of settlement and crude agriculture. The settler brought with him grain fields, civilized weeds, and rail fences. He converted an increasing area of woods into brushy stump lots, and on the prairie he added osage hedges to the quail environment. Grain, brush, weeds, and hedges stabilized the quail crop and extended the area of quail range. Clearings extended it further into the woods; hedges extended it further into the prairies; grain extended it northward into new latitudes. At the same time all these changes probably also increased the per acre population. Quail were sporadically trapped, but never shot in the early days. There was a gradual but very large increase in total numbers and in distribution, which lasted, in the greater part of this region, for more than fifty years.
>
> Third came the era of agricultural intensification. The weedy rail fence was replaced by naked wire. Brushy woods were converted into bare pasture, and hedges were uprooted from the prairie farms. With these unfavorable changes came also bird dogs and shotguns, cheap ammunition, and increasing leisure. These processes of modernization began early on the prairies, but only a decade ago in the Ozark hinterlands. They were accompanied by a decrease in quail, frequently due to overshooting, and nearly always due to a decrease in the area of habitable range.

The "fourth stage" depicted by Leopold concerned the effects of the agricultural depression of the 1930's, which need not concern us here.

The peak of the California Quail population, occurring in the period 1860 to 1895, would correspond very well to "the era of settlement and crude agriculture," which in California also should include "the era of crude pastoralism." The initial widespread ecologic impact on the landscape of California was triggered by the development of immense herds of livestock. Dana and Krueger (1958:16) state:

> Following secularization of the missions in 1833, came the pastoral era when agriculture was dominated by the great cattle ranches. The rapid growth in population that followed the gold rush and California's admission to the Union resulted in an impressive expansion of the livestock industry. During the ten years following 1850 cattle increased to about 1,000,000 head and sheep to perhaps twice that number. Both cattle and sheep were hard hit by the great drought of 1862, but sheep made the better recovery and by 1876 reached their all-time high, with an estimated 7,700,000 head.

Concurrently, the better soils were being broken by the plow and planted to grains, hay, orchards, and vineyards. The first major cash crop was barley, but by 1860 wheat had become dominant with a yield of nearly 6,000,000 bushels. Grapes and fruit trees were added with the advent of irrigation systems.

Picture the landscape in a typical valley of the inner coastal ranges of central California about 1870. The oaks and sycamores along the stream course, tangled with wild grape vines and undergrown with shrubs, had not yet been disturbed. But the rich alluvium on benches bordering the stream were plowed in small patches, and the more accessible level areas were planted to grain.

A few small orchards and vineyards had been set out in areas easily irrigated from the stream. The grassy hills were being heavily grazed, breaking up the native bunch-grasses and allowing an intrusion of many seedbearing Mediterranean forbs such as filaree and clovers. These same forbs densely invaded the poorly kept orchards and vineyards and likewise sprinkled the grain stands. One could scarcely envision a better habitat for the quail. Food, cover, and water, occurring in abundance on soils not yet depleted of their virgin fertility, would stimulate production of a great flush of seed-eating birds, such as quail. Some such ecologic circumstance unquestionably underlay the abundance of quail in the early years of settlement.

A point that needs great emphasis is that the fortuitous production of optimum vegetation for quail took place on soils brimming with the stored fertility and organic matter of the ages. The same was true of the peak period of Bobwhite production in the Midwest. It is unrealistic to believe that these pioneer conditions could be fully restored today by proper land management. Overgrazing, overcropping, and surface erosion have stripped

Figure 10. A remnant of the original bottomland quail habitat along the Sacramento River, Bidwell Park, Chico. J. B. Cowan photo.

most lands of that accumulated richness that came with centuries of soil maturation under native vegetation. Perhaps only the deep alluvial valleys have retained the basic capacity to fully renew their original productivity, and those are the areas now cultivated most intensively and mechanically. We must take the sensible view that the great quail peak of the late 1800's is a glamorous relic of the past—a relic we wish fully to understand but that we can never actually reproduce.

By the same token, a virgin forest initially logged may produce a rich second-growth of brush and tree reproduction that in turn may support a fabulous deer population. No subsequent cutting of second-growth timber produces quite the same profusion of animal life. A newly filled reservoir often supports a tremendous fishery, but as the organic matter and available nutrients of the bottom are drawn away, productivity of the basic food chains slows down and the fishery dwindles.

Some such ecological sequence must be envisioned to explain the great crest of the quail population in the latter half of the 19th century.

THE QUAIL DECLINE

And what of the decline? According to Walter R. Welch (quoted above), the decrease in the quail population had begun by 1895, and the era of general abundance apparently was over by 1925. What events, singly or in combination, could account for this dramatic collapse in so short a period?

It was natural that contemporary observers placed the blame on obvious sources of mortality—hunting and predation primarily. Particular acrimony was directed toward the market hunter and illegal poacher. Grinnell et al. (1918:536) summarized this widely held view as follows: "Hunting in one form or another has been the most effective factor in the decrease of the Valley Quail. As with so many of our other game birds, too long an open season, too large bag limits, or none at all, and hunting for the market, have together been instrumental in reducing the numbers of quail; but the last-named factor is undoubtedly the most important one."

Many anonymous notes and announcements in *California Fish and Game* emphasized the need for better protection from overhunting. A notice appearing in 1923 (p. 109) stated that objection was being raised to the year-round open season on rabbits in southern California because this served as a screen for the nefarious activities of quail poachers. In 1925 (p. 74), it was noted that some counties (Monterey and San Benito) were asking for a closed season on quail. In that same year (p. 76), a cryptic note detailed how in the 1890's local ranchers near San Diego had habitually invited friends to shoot quail that were allegedly damaging their crops, and at the time of writing, 40 years later, there were "practically no quail" in evidence. In each case, the decrease was ascribed to hunting.

The seven ex-market hunters interviewed by Twining (1939) admitted that past overshooting had been deleterious to quail numbers, although they felt that other factors such as predators, broadcast distribution of poison baits for ground squirrels, and decreased availability of water were ancillary influences. Welch (1928), viewing the problem from the vantage point of time, offered the more balanced opinion that over-hunting plus habitat destruction by agriculture and grazing had led to the decrease.

E. B. Ralston (1916:188) recounted his own exploits as a market hunter, but blamed the decrease in quail on changing land use. He noted that when settlers "began grubbing and clearing the hills and flats, the quail began to lessen in numbers. They were thus driven out of their feeding and watering places, away from cover and protection, and cattle and horses, in feeding on the wild grass, exposed and destroyed their nesting places, contributing largely to the decrease. . . ." Perhaps Mr. Ralston was closer than most of his contemporaries to the actual explanation.

In my appraisal, what appears to outweigh all other possible causes of this historic quail decline is a decrease in production of quail food that undoubtedly occurred in the same period. This aspect of the problem is notable for its lack of mention in the earlier studies.

I have outlined the correlation between the fabulous quail abundance of the period 1860 to 1895 and the concurrent disruption of aboriginal range conditions through introduction of European agriculture with intensive grazing and the advent of the alien annual plants, such as the filarees and weedy legumes, that produced prodigious supplies of new quail food. Flourishing for a few decades on the stored fertility of virgin soils, the new production was abnormal and above what could be sustained. Inevitably it would decline and so would the dependent quail population, all to a great extent attributable to continued land depletion.

However, along with the soil depletion, and as part of the continuing ecologic change and disruption, new plant species mostly of Mediterranean origin continued to invade the California Quail country. Accompanying or following the alien filarees and legumes came various annual grass species, mainly wild oats, bromes, and foxtails. Well adapted to California's Mediterranean climate, highly competitive but of little value as quail food, the new annual grasses have steadily gained dominance on many ranges, commonly to the point of establishing solid stands. Where this occurred, the broad-leaved plants, both alien and native, that were then and still remain the main producers of quail food, were excluded or greatly reduced. Probably the most significant aspect of the quail decline is the correlated replacement of annual forbs that produce abundant quail food, by annual grasses that do not.

3

EARLY EFFORTS TO RESTORE QUAIL IN CALIFORNIA

THE GENESIS OF MANAGEMENT

Wildlife conservation in the United States became a serious enterprise about the turn of the century. Prior to that time, protective game laws had been passed in all of the states, and, in fact, in twelve of the original thirteen colonies prior to the Revolution; but there was little concerted effort to enforce them. By the late 1890's, the easily observable decrease of many game species became a matter of growing concern to hunters in particular and to some legislators as well, setting the stage for the "conservation era" dramatically introduced and sponsored by President Theodore Roosevelt. Following the conference of State and Territorial Governors, called by the President at the White House in May, 1908, there was a rash of activity in creating and strengthening state conservation agencies of many kinds, including fish and game departments.

As regards the endeavor to protect and restore species of wild game, there ensued an orderly sequence of procedural steps and ideas that have been categorized by A. Leopold (1933). The history of efforts to restore the California Quail followed this classical sequence almost to the letter.

1. *Protection*.—The view was widely held that overhunting was the major cause of game shortage, so protective laws were adopted and

wardens were employed to enforce them. Regulation of the kill has throughout history been the first step in a wildlife conservation program.

2. *Refuges.*—Establishment of inviolate sanctuaries was a corollary form of protecting game from hunters.
3. *Predator control.*—Further protection of game breeding stocks was attempted by directing the energies of departmental employees and the general public alike toward the killing of predatory animals.
4. *Restocking.*—Breeding stocks were augmented by releasing birds or mammals raised in pens, or wild-trapped and imported from areas of abundance.
5. *Habitat improvement.*—Always last in the sequence has been the endeavor to improve the environment for game so that it could better care for itself.

PROTECTIVE REGULATIONS

As noted in the previous chapter, the California legislature in 1880 outlawed the commercial trapping of quail, and in 1901 the sale of California Quail on the market, however obtained, was declared illegal. Although these legal steps acted as a deterrent to market hunting, they did not stop it completely for at least two decades. The market price for quail rose substantially, encouraging poaching and subversion. However, by the mid-1920's commercial hunting and trapping had ceased. This result was hastened by the dramatic decrease in quail numbers, as already recounted.

Sport hunting for California Quail has been regulated by law since the turn of the century, allowable bag limits and hunting seasons varying from year to year and from locality to locality. In 1931, the season was from November 1 to December 31 in part of the state, November 15 to December 15 in the balance. The daily bag limit was 15, with a weekly bag limit of 30 (two days allowable kill). In recent years, the season has been somewhat lengthened to 2 to 4 months in different districts of the state, and daily bag limits have varied locally from 6 to 10 birds.

Enforcement of the regulations has been generally adequate, but legal protection alone has not stemmed the quail decrease, for the reason that hunting is no longer a limiting factor on quail numbers, if indeed it ever was. When a game species decreases in abundance, it is easy to ascribe the change to the kill by hunters. In the case of some big game species such as deer or elk, this indeed was an important factor historically. But small game populations, with a high reproductive rate like quail, can absorb substantial removal by hunting without materially influencing the basic, year-to-year level of numbers, which more often than not is determined by the adequacy of the habitat in which the population lives. Regulation of hunting in no way can compensate for habitat deterioration.

Protective regulations, therefore, however carefully they were designed and enforced, did not effectively restore the waning numbers of California Quail. Further comment on hunting regulation is offered in Chapter 14.

QUAIL REFUGES

In the 1920's and 1930's, a plan was developed to restore quail by establishing refuges in widely scattered localities over California through cooperative agreement between private landowners and the Fish and Game Commission. Welch (1931b:424) describes the program as follows: "This campaign . . . has met with the spontaneous cooperation and support of farmers, ranchers and landowners throughout the State, with the result that during the past year upwards of 1000 inviolate quail sanctuaries and game refuges have been established. . . . These quail sanctuaries . . . are located in nearly every county of the State from Modoc to San Diego, and contain about 600,000 acres of land, upon which it is estimated there are at this time about 300,000 wild Valley and Mountain quail." Each participating landowner signed a "gentleman's agreement" to set aside part of his property as a sanctuary, whereupon he was supplied with signs to post the closure and assistance from cooperating sportsmen and game deputies in patroling it. The signed agreement was for a period of three years, subject to renewal. In 1931, Welch reported the number of quail refuges had increased to 1200.

The refuge program did little to overcome the basic quail shortage, which had resulted more from habitat deterioration than from overhunting. Within a few years the refuge effort was abandoned.

PREDATOR CONTROL

Concurrent with the refuge program, a concerted effort was made to reduce the number of natural enemies of quail, particularly on the designated sanctuaries. Welch (1931b:425) summarized the endeavor for one six-month period as follows:

> In conducting their work of predatory bird and mammal control on quail sanctuaries and game refuges, the volunteer deputies of the Division of Fish and Game, between January 1 and July 1, 1931, have destroyed the following number of birds and mammals that are considered to be enemies of quail: 38 coyotes, 33 bobcats, 684 house cats, 35 foxes, 43 coons, 8 weasels, 2 opossums, 1 badger, 5 wild and unclaimed dogs found destroying deer and the nests of quail, 365 sharp-shinned and cooper hawks, 3972 blue jays, 293 magpies, 81 crows, 49 butcher birds, 2 great horned owls, and 47 snakes.

To supplement this enthusiastic across-the-board predator campaign by volunteer deputies, the Division of Fish and Game built up a considerable force of animal trappers who used traps, guns, and dogs in an all-out war against the natural enemies of game. This force reached 45 full-time men in 1948. Over the years, it became apparent that the costly war against predators was having little beneficial effect on most game populations. Concurrently, research on quail and other game species was demonstrating that the real controls over numbers lay in deficiencies of the environment, not in the direct sources of mortality. So, in 1957, the force of predator trappers was disbanded, and the men were assigned other duties in the Division. The futility of general predator control as a measure of quail management is now firmly established.

ARTIFICIAL RESTOCKING

Another management measure, tested on the quail refuges, was the artificial replenishment of existing breeding stocks by birds reared in game farms or imported from areas where quail were abundant. The Los Serranos Game Farm devoted to quail propagation was established at Chino. True (1933a) states that quantity production got underway in 1933 with the rearing of 4000 birds which were liberated on established refuges in southern California. The following year, the restocking program was augmented by the importation of wild-trapped California Quail from Baja California, as already recounted in Chapter 1. Between 1933 and 1941, 65,000 quail, derived from the game farms and from imports, were liberated in southern California. But as a management technique the "quail replenishment" program proved to be expensive and ineffective. According to Welch (1931b:422), hand-reared quail cost from $2 to $7 per bird. While the wild-trapped birds from Mexico were very much less expensive (True, 1934),

Figure 11. Rearing pens at Los Serranos Game Farm, Chino, where California Quail were propagated for release. California Department of Fish and Game photo.

the ultimate effect of adding them to existing wild stocks was not evidenced by any sustained increase in local populations.

All liberated birds were banded, and some interesting points were derived from analysis of the band returns. Richardson (1941) points out that mortality in a cohort of quail liberated in hunting territory (almost entirely wild-trapped birds) was about the same as that computed for native wild birds by Sumner (1935:247):

	Alive at the end of:			
	1st year	*2nd year*	*3rd year*	*4th year*
100 liberated quail (Richardson, 1941)	27.8	7.0	3.1	0.7
100 native quail (Sumner, 1935)	26.8	7.2	2.0	0.5

If the liberated wild-trapped stock was unadapted, there certainly was no indication in poorer survival of adult birds. On the other hand, nothing was learned about the comparative success of the two stocks in rearing young. In the midwestern United States, there was a period in the 1920's and 1930's when enormous numbers of Mexican Bobwhites were imported for restocking in Louisiana, Arkansas, Texas, and some other states (Rosene, 1969:21). Again the results were poorly documented, but the general consensus seems to have been that the experiment was a failure, for imports were terminated in the late 1940's.

The California Department of Fish and Game no longer attempts to increase quail by a general restocking program, although local transfer of breeding stocks may be justified when a population has been depleted by

Figure 12. Quail breeding stock in a trap, Los Serranos Game Farm. California Department of Fish and Game photo.

some natural or man-caused catastrophe. In the Great Basin, periodic winter-kill of quail dictates the need for reintroduction of breeders from time to time.

HABITAT IMPROVEMENT

Concurrent with all the above endeavors made on behalf of the quail population, it was well recognized that the birds could only thrive in proper habitat that included necessary types of cover, food, and drinking water. On a practical scale, there was very little that the Division of Fish and Game could do about supplying cover and food. In the quail refuges where birds were being released, some brush piles were built to augment natural cover, and grain was distributed periodically to hold the liberated birds in the refuges and keep them from wandering (True, 1933a). But it did prove practical to add water sources where other aspects of habitat were favorable and water was lacking. The initial watering device was a 55-gallon drum situated in a desirable location and connected with a shallow drinking trough (True, 1933a). The Division of Fish and Game installed many such drums in the designated quail refuges, and the Forest Service adopted this management strategy on the San Bernardino National Forest for the encouragement of Mountain Quail (Rahm, 1938).

The problem with water drums, however, is that they require periodic filling. An extensive system of such devices would entail a substantial program of maintenance. In 1943, Ben Glading and Fred Ross invented a self-filling watering device which came to be known facetiously as "Glading's gallinaceous guzzler." The guzzler consists of an underground tank, made of concrete or pre-formed plastic, which receives the rain water that falls on an adjoining oiled apron. Quail reach the underground water via a ramp. Properly built, a guzzler may last almost indefinitely with a minimum of maintenance. Here was a practical management device that could be used to extend quail habitat into great areas of arid range in southern California. Details of guzzler construction and use will be described in Chapter 13.

Between 1943 and 1952, the Department of Fish and Game installed 1,921 guzzlers in California, mostly in the southern counties. Macgregor (1953) reported on an evaluation of the guzzler program and generally found that it had been effective in developing quail populations where none existed before in arid and semi-arid ranges that were lightly stocked with cattle. In areas that had been over-grazed or otherwise rendered deficient in food and cover, the addition of a water source was ineffective in restoring breeding stocks. The guzzler program nevertheless was continued, and by 1972 a total of 2,150 had been installed by the Department, plus many others privately constructed.

It was soon recognized that the overall management of quail habitat was a form of land husbandry that could be practiced only by the landowner. Preservation of coverts and minimizing the adverse effects of agricultural

Figure 13. Ian McMillan (left) explaining his program of brush restoration for quail cover to Richard Genelly. Only the land operator can effectively manage quail habitat.

practices involved day-to-day decisions made in ranch operations. Emlen and Glading (1945) prepared a bulletin to instruct landowners in the basic habitat needs of quail and ways in which these could be met. *Increasing Valley Quail in California* was widely distributed by the Cooperative Extension Service and represented a conscientious effort to bring about improved conditions for quail through education. Doubtless this bulletin had some salutary effects, but it did not stem the continuing deterioration of habitat that has been an outgrowth of the agro-revolution.

RÉSUMÉ OF MANAGEMENT EXPERIENCE

The history of management effort on behalf of the California Quail supports the conclusion reached by wildlife biologists regarding the husbandry of any wild species—namely, that the maintenance of adequate habitat is always the key to success. The primary factor that will determine the status of a quail population is the suitability of the environment in which it lives. Protection from overshooting is assured nowadays by legal regulation of hunting, yet it is ineffective in areas deficient in food, cover or water. Still less profitable is the endeavor to increase populations by wholesale killing of natural enemies or by the augmentation of wild stocks with liberated birds.

The California Department of Fish and Game found one practical way in which it could improve quail range by adding sources of drinking water where that was in short supply. But the equally necessary maintenance of food and cover resources is beyond the purview of a regulatory government agency. Only the owner and operator of the land, private or public, is in a position to make decisions about land use that will determine the suitability of vegetation needed by quail.

There is a paradox in the legal basis for conservation of a resident game species such as the California Quail. In the State is vested ownership of the game and responsibility for its management. Yet the State is largely powerless in exercising the on-the-ground husbandry required to maintain a suitable habitat where the species can exist. Some public agencies such as the Forest Service, the Bureau of Land Management, and the military services manage large blocks of public domain land that include fine areas of California Quail range capable of sustaining good populations of the species. Management decisions arrived at by these agencies (i.e., grazing intensity, brush eradication, water development, etc.) will greatly influence the welfare of quail. Private lands, of course, are used at the judgement and discretion of the landowner, whose decisions likewise may be favorable or deleterious to quail. In this process of land-use decision making, the Department of Fish and Game plays a minor role or none at all. Yet the Department is nominally the custodian of wildlife.

If this volume on the California Quail is to serve any real purpose, it will be through bringing to the attention of landowners and operators—public as well as private—the rudiments of management of the landscape necessary to favor the welfare of the species. Past efforts by the Department of Fish and Game to restore quail populations by regulating hunting, establishing refuges, controlling predators, and artificial restocking have fallen far short of the mark. The only approach to quail management that offers assurance of success is habitat restoration and maintenance by those individuals and agencies that manage land.

4

LAND USE AND QUAIL HABITAT

CLEAR THE VALLEYS, GRAZE THE HILLS

The very influences of agriculture and pastoralism, which in their initial form created optimum quail habitat, led to habitat depletion as they were intensified. Small fields tilled by horse power were combined into larger and larger fields as the farms became mechanized. The sheltering fence-rows have vanished. Brushlands have been cleared for pasture improvement. Hill lands continuously grazed for a century have lost fertility and productivity. Many hilly areas, once cropped for grain, have reverted to pasture with much less food value for quail. Stream bottoms, fence-rows, and odd corners have been cleared for the sake of agricultural efficiency. Water tables have dropped in many parts of California, with a consequent loss of watering sites for quail. Weedy roadsides have been systematically mowed, sprayed, or burned to control weeds. The trend toward slick and clean agriculture has had cumulative adverse effects on the quail habitat.

By and large, the California Quail no longer occurs abundantly in the valleys of California. Intensive use of the deep alluvial soils, which once supported maximum quail populations, has essentially removed the cover required by the bird in day-to-day living. As late as the 1930's, there remained scattered patches of uncleared and undeveloped bottomlands in the Central Valley. One of these remnants on the Parrott Ranch, 6 miles west of Chico, Butte County, retained dense populations of Black-tailed Deer and California Quail in the tangled grape vines and riverine oaks that survived along the banks of the Sacramento River. In 1933, the Museum of Vertebrate Zoology sent a collecting party into a block of 8000 acres of native oak forest on the Parrott Ranch. The field notes of A. H. Miller and D. S. McKaye describe graphically the open stands of valley oaks, draped with wild grape tangles in moist sites. McKaye wrote (Sept. 18, 1933): "Quail are here by the thousands." Shortly thereafter, the timber was cut and the land cleared for agriculture. This process has been nearly universal in rich valleys, large and small, throughout California. Quail populations have been reduced to remnant coveys, persisting around homesteads, suburbs, railroad rights-of-way, levees, and other areas retaining bits of shrubby cover. Only in the states to the north and east of California are valley lands still widely habitable for quail.

The "valley quail" today is, in effect, a bird of the hills. There the major adverse range effect has come from excessive grazing. It is true that light grazing is advantageous to quail in some situations, where it breaks the density of grassy cover and favors the growth of important quail foods such as lupine, lotus, filaree, and bur clover. But overgrazing goes the next step, leaving range that is cover- and food-deficient and hence of little use to quail. More importantly, overgrazing depletes the fertility of the thin soils, lowering both quality and production of quail food. As implied above, the fertility of the depleted hill lands cannot easily be restored.

It is not to be expected that the clock of California history can be turned back to the era of pioneer agriculture, even locally. Neither, however, can heedless loss or damage to the California rangelands be accepted any longer as a prerogative of short-term agricultural exploitation. Perhaps the basic issue is not what man is doing to quail habitat but rather what he is doing to his own habitat. The extirpation of the California Quail from its native rangelands is evidence of land misuse that in the long run will be reflected in lower yields of livestock and other economic products. Wildlife abundance or scarcity is a sort of barometer of the "ecologic health" of a landscape.

REGIONAL CHARACTERISTICS OF QUAIL RANGE

The range of the California Quail in western North America is so extensive and ecologically so diverse that I find it difficult to make categorical statements about habitat needs or habitat changes affecting the species as a whole. To facilitate discussion of regional problems of land use and quail natural history, I have divided the total present range into four major districts or ecologic subdivisions of the range. These zones and some of their characteristics are depicted in Figure 14 and in the descriptions which follow.

1. *Arid ranges*. Most of the original range of the species falls in the arid zone, which includes the San Joaquin Valley and foothills, the coastal ranges south of Monterey Bay, the foothills of the southern Sierra Nevada, and deserts and brushlands southward to the tip of Baja California. This zone is characterized by low rainfall, sparse ground vegetation, and mild winters. Lack of cover is often a limiting factor for

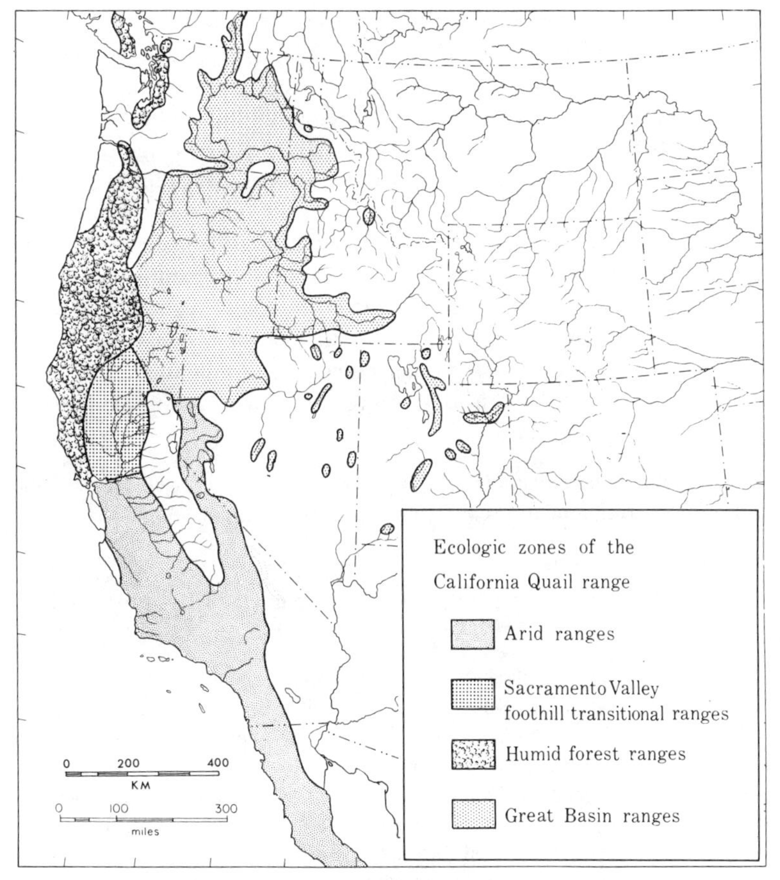

Figure 14. Ecologic zones of the California Quail range.

quail. Overgrazing by livestock as it affects quail food supply is a major land use problem. Years of high rainfall are favorable for quail reproduction. A sequence of drought years may reduce the quail population to a very low level.

2. *Sacramento Valley transitional ranges*. The foothills surrounding the Sacramento Valley constitute the most stable quail habitat in the whole range of the species. Rainfall is moderate, and annual fluctuations in precipitation do not drastically affect the quail population. Ground vegetation is moderately dense, cover is generally adequate, and some grazing by livestock is often beneficial to quail welfare. Winters are mild and snow is rarely persistent.
3. *Humid forest ranges*. The coastal forest zone is characterized by a heavy forest canopy in many areas and dense ground vegetation where the forest is thin or absent. By and large, the quail thrive only in localities where the ground has been cleared by logging, fire, agriculture,

grazing, or some combination of these. Much of this zone was only lightly populated with quail, or completely unpopulated, until the country was settled and developed for assorted uses. Dry years are favorable for quail reproduction, wet years are highly unfavorable.

4. *Great Basin ranges.* Most of the newly occupied range of the California Quail lies in this zone, which includes the Columbia drainage and scattered river valleys throughout the Great Basin. Ranching operations converted much of this region into habitable quail range, through irrigation, agriculture, and the establishment of farm yards where quail can retreat in hard winters. Ground vegetation is sparse, and overgrazing is commonly a limiting factor. Deep snow and cold cause periodic die-offs of quail populations, which recover in years of mild weather. Warm dry springs are favorable for reproduction, cold wet springs are unfavorable.

In view of the substantial ecological differences between these zones, I shall make frequent reference to them in the chapters that follow.

BASIC HABITAT NEEDS

Emlen and Glading (1945:5) characterize the habitat needs of the California Quail as follows:

> The carrying capacity of a tract for quail is determined by the presence and distribution of food, water, and cover in acceptable forms. Moving about mostly on foot, quail require a broken terrain with a close intermixture of open feeding areas and protective brushy cover. Any farming operations that create broad expanses of single vegetation types are not favorable to quail, even when such areas consist of valuable food or cover plants. A clean-cultivated vineyard, for example, has the necessary brushy element, but lacks food; a large grain field contains much good quail food, but lacks brushy cover. Such large areas of any single cover or food plant are of little or no use to quail except along the edges. Mixed vegetation with small clumps of trees and shrubs scattered among open feeding areas, as on many foothill range lands, provides a maximum of "edge" and is ideal for quail when water also is available.
>
> The essence of habitat management for quail is the creation of the best possible mingling of feeding areas, protective cover, and watering sites.

This statement of quail habitat requirements, written over 30 years ago, is completely acceptable today. Continuing study and experience in management have elaborated many details of quail ecology, but the concept of habitat dependency has not changed.

COVER

The term "cover" is used to designate the various forms of vegetation, living or dead, that quail utilize for shelter from the elements or from natural enemies. Most forms of quail cover consist of living woody plants such as shrubs, briars, and dense trees. But the birds are quick to take advantage

Figure 15. Chamise and oak cover along a dry wash near Dunnigan, Yolo County, showing good interspersion of brushy shelter and open feeding areas. Several quail coveys frequent this area.

of any structure that offers protection when it is needed. I once saw a quail escape the close pursuit of a Cooper Hawk by flying at full speed through a 2-inch mesh chicken wire—a risky feat at best, but successful in this case with only the loss of some expendable plumage. Dead vegetation in the form of brushpiles, windrows of tumble weeds along a fence, or dried stands of weed stems are frequently used. But living brushy vegetation is usually the basis for quail occupancy.

Brush cover serves primarily as the daytime "home" of a quail covey. After the morning feeding, the birds repair to a safe and comfortable haven sufficiently dense to supply, as required, shade from the sun, shelter from rain or snow, and escape from the attacks of hawks or other predators. There they loaf and spend the day at leisure until it is time for the evening period of feeding. I know of very few places where California Quail exist without some living shrub cover. The actual acreage of such cover need not be extensive, but the density is important. Unless the birds have a secure refuge where they can rest in safety during the day, they probably will not settle. Daytime loafing cover can perhaps be designated as the key building block of quail habitat.

Food is another basic requirement of quail survival, and it is essential not only that the supply exist but that it be safely accessible. A principal role of shrub cover as a component of quail habitat is to enable the birds to reach food that would otherwise not be available. In decades of hunting quail each fall and winter on extensive ranges in central California, I repeatedly have found the respective coveys practically in the same locations year after year. The particular coverts preferred by the birds are of the same general type and density as the shrub cover on unoccupied parts of the same range. What has finally been noted is that the preferred coverts are consistently adjoining or interspersed with areas productive of quail food.

Ian McMillan has pointed out to me that the shrubland communities that are the characteristic habitat of the California Quail are often found on soils that are naturally suited to the growth of seed-producing forbs. Commonly, areas of these lighter, more delicate soils lie contiguous to or intermingled with heavier, more stable grassland types. In some years, the grassland soils produce quail food in abundance, as also do some of the brushland areas. Little is known about the extent to which each type may produce the most or best quail food. Certainly the food on the shrubby areas is most available. Some of the highest and most stable quail populations are found in ''edge'' situations of this kind where quail food on the open areas is made available by the contiguous shrubland growth. Sparse brushlands interspersed with good forb growth represent the optimum mix of cover and food for quail.

There are some situations where rough broken outcrops of rock serve as adequate escape cover for quail. At various points along the western Sierra foothills, granite or basaltic outcrops are a major form of quail cover. The same situation applies in the rimrock country of northern California, eastern Oregon and Washington, and along the Snake River in Idaho. The immature soils in and around outcropping rock are often suitable for the growth of seed-producing forbs which are protected from overgrazing by the very irregularity of the ground surface.

At night, California Quail prefer to roost off the ground, normally in dense trees or tall shrubs. This is not true of all quail. The Bobwhite, for example, roosts on the ground, a covey forming a tight circle, tails together. Many gallinaceous birds, like the pheasant and some grouse species, roost singly in tall grass or protective dead weeds. But California Quail use communal roosts above the reach of ground predators. A juniper tree, a dense live oak, a bay tree, or even a tall atriplex bush serves the purpose admirably. There are extensive areas of potential quail range in the arid southern parts of California that are decidedly lacking in roosting cover. This fault can be rectified by planting appropriate roosting trees or by building artificial roosting structures—elevated brush piles—in a manner to be discussed in Chapter 11. But an adequate roost is an essential item on the list of requirements for quail occupancy.

Quail are quite adaptable in the matter of finding proper shelter for nesting. The birds prefer to nest in rather open situations, on the fringe of a clump of low brush or tall grass, apart from areas of continuous overhead cover such as chaparral or forest. Shade from the sun and concealment from predators are two requisites of good nesting cover. Hilly pasturelands, lightly sprinkled with shrubs, are often used by nesting quail. However, heavy grazing may remove all the annual growth under the canopy of the shrubs, rendering an area unsuitable for nesting. Fence-rows and narrow gullies offer good nest sites, but these strips serve as "highways" for ground predators such as skunks, raccoons, and stray cats, that find and destroy a considerable number of quail nests so situated. Crows may use fence-posts for observation perches, thereby discovering quail nests below. By and large, nesting is most successful when the individual nests are randomly scattered over lightly grazed pasturelands.

FOOD

The basic component of the quail diet is the assortment of seeds produced by various species of broad-leafed annual plants, botanically termed forbs. Depending on geographic area, examples of the principal quail food producers in California are lupine *(Lupinus* sp.*),* lotus *(Lotus* sp.*),* filaree *(Erodium* sp.*),* clover *(Trifolium* sp.*),* bur clover *(Medicago* sp.*),* fiddleneck *(Amsinckia* sp.*),* and similar broad-leafed plants. If available, quail make good use of acorns, waste grain in stubble fields, and fruits and berries, both wild and cultivated. But the staple foods that support the vast majority of California Quail through the year are the seeds of wild forbs, with emphasis on the legumes.

In areas of moderate to high rainfall, grass species such as wild oats *(Avena fatua)* tend to dominate ground cover to the exclusion of broad-leafed annuals. Dense grass stands produce little quail food and may form a sod that, unless disturbed, changes little from year to year. In the coast ranges and the more mesic Sierra foothills, thinning of the grass cover by grazing or other means may be required to create some bare ground where forbs can grow and where quail can forage freely.

I hunted quail for some years on the Penobscot Ranch near Georgetown, in Eldorado County. A pasture that initially had an excellent quail population was withdrawn from grazing in 1970, and the number of birds dropped perceptibly in the next three years, although on other parts of the ranch there was no apparent change in population levels. A dense stand of grass (mostly *Poa*) enveloped the ungrazed pasture to the extent that seed-bearing forbs were much reduced and only a few small coveys of quail persisted. Emlen and Glading (1945) recognized that undergrazing in mesic areas sometimes has an adverse effect on quail range.

On arid ranges where rainfall varies extremely from year to year and is not sufficient to support a stable grassland community, forb growth occurs more commonly and naturally than in the more humid parts of the quail range. Areas that may become bare of annual growth in dry years can produce abundant supplies of quail food in wet years. The resident quail populations fluctuate accordingly. However, even in these arid sections, as throughout the entire range of the California Quail, production of quail food depends mainly on the extent to which broad-leafed annuals are able to outgrow the annual grasses.

Natural factors other than rainfall strongly influence the production of quail food. Vegetation on south-facing slopes commonly differs from that on the north slopes of the same ridges and hills. Grasses do best on the shady north slopes, forbs thrive better on the sunny, drier, southern exposures. Some of the steeper south-facing slopes of arid inland ranges in central California are perennially in unstable conditions that may favor the growth of forbs. This is partly due to continual downhill movement of the surface soil, which in some ranges may be high in plant nutrients. In wet years, south slopes of this kind become the favorite foraging areas for the resident quail.

Various species of wildlife, including quail, importantly influence plant competition and thus the production of quail food. Kangaroo rats *(Dipodomys),* pocket gophers *(Thomomys),* and field mice *(Peromyscus),* by reason of their burrowing habits, are important agents of soil formation and plant succession (Grinnell, 1923). Rodent populations fluctuate widely and reach high periodic densities, particularly in arid sections of the quail range. In time of peak numbers, kangaroo rats may remove most of the annual growth on some areas, and their trails and burrows affect the land in various ways to change the composition of annual growth. Abundant forb growth commonly occurs on areas that were previously perforated, aerated, and cleared of vegetation by kangaroo rats.

The main supplies of fall and winter quail food are seeds that lie in range litter. Held from direct contact with the mineral soil, these seeds may lie unsprouted for years, to be available as crucial food reserves in years of drought or other causes of failure in annual seed production. In their characteristic foraging behavior, quail get these seeds by scratching and turning over whatever light duff or plant residue may cover the soil surface. These workings often cover the entire surface of extensive areas. Scratching and winnowing of the surface material has the effect of light tillage and may favor the germination and growth of forbs over grasses. Conversely, in selecting the seeds of their principal food species, and in grazing the new sprouts and foliage of these same species, quail in abnormally high density can adversely affect their own future food supply. In the immediate environs of the Ian McMillan homestead, intense foraging by 1,000 or more quail has greatly thinned the growth of forbs, forcing the birds to forage several hundred yards from protective cover and exposing them to predation by the wintering hawks (Fig. 16). Under normal circumstances, the natural regulation of quail numbers may serve to prevent densities that could adversely affect the production of quail food.

Figure 16. Shrub cover around the Ian McMillan house is the nucleus for the "home covey" of over 1,000 quail. The bare ground is caused by quail consuming all green forbs and grasses growing near the cover. M. Erwin photo.

Arid ranges usually offer plenty of bare ground, but excessive grazing by livestock thins both grasses and forbs, often precluding the set of seeds. Filaree and various annual legumes may sprout abundantly, but if closely cropped the plants produce little seed. Surface duff in which seeds remain unsprouted as a reserve supply of quail food does not accumulate on overgrazed pastures. Additionally, overgrazing may deplete or eliminate the shrub cover needed by quail.

With the onset of winter rains, some of the seeds in the soil sprout and a new source of food becomes available in the form of succulent greens. The birds apparently welcome this dietary supplement, since greens show up in quail crops shortly after the rains begin. Most of the greens are juvenile plants of the same seeds that constitute the dry season diet; hence the feeding grounds may be more or less the same through much of the year. Greens alone, however, will not sustain quail in good health. A proper winter and spring diet should contain a generous component of seeds with high caloric content. Moreover, greens vary from year to year in nutritive value, as will be elaborated in Chapter 9.

Over the years, the Food Habits Section of the Wildlife Investigations Laboratory, California Department of Fish and Game, has analyzed the contents of 2,352 quail crops, taken in many parts of California at all seasons of the year. Bruce Browning of that Laboratory has compiled and analyzed these data for publication as an original research report in Appendix B of this volume. For a full discussion of the food habits of California Quail, reference is made to the Appendix. Additional discussion of the significance of the quail diet and ways to supply it will be forthcoming in Chapter 12.

WATER

The third requisite of quail range is drinking water, situated close enough to the cover and feeding areas to be readily available. Water is little used throughout most of the year, but is critical in summer and early autumn, when temperatures are high and young quail are being reared. Any little drip or puddle will serve, so long as it is dependable. In aboriginal California, the valleys generally had plenty of permanent water for quail, but the foothills and arid rangelands offered only those springs and seeps that nature had supplied. With settlement, the water regime in California has been drastically altered. Pumping from deep wells has lowered the water table, and many original seeps have gone dry. On the other hand, irrigation projects have brought water to many areas that were formerly dry. Unfortunately, the irrigated farmlands tend to be intensively cultivated, so that cover for quail is largely eliminated; hence the water serves little purpose in supporting quail. Windmills, stock watering troughs, and ponds have been widely installed on grazing lands, but characteristically the food and even the cover plants are largely consumed by livestock lingering near the water; hence the sites are not attractive to quail. The general water situation for quail has probably deteriorated in most quail ranges of the state, compensated only in part by the program of guzzler construction.

Figure 17. A stock watering trough in arid San Luis Obispo County, used by a large covey of quail. Nearby cover makes this water safely available for the birds. M. Erwin photo.

In summary, habitable quail range is a mosaic of cover types, food sources, and watering points. Every change in land ownership or in land use portends a change in the suitability of the landscape to support quail. Some of the changes that have occurred in the last century are favorable, but most have been adverse.

CHANGES IN CROPS, TILLAGE, AND GRAZING PRACTICE IN CALIFORNIA

The 12 million acres of superior tillable land in California have undergone a sweeping alteration. A substantial area has been rendered unavailable for wildlife use by urbanization, highway construction, and other forms of development. But the bulk of the area is intensively cultivated for crops, orchards, and vineyards, to a degree that has largely eliminated the cover that quail require. The single family farm is disappearing in favor of large holdings that are operated as corporate agro-businesses. Fences are removed and fields are greatly enlarged to facilitate mechanical tillage. A substantial and growing portion of the crop land is being irrigated to augment yields; the high cost of irrigation demands intensive utilization of every acre to maximize production. Herbicides are used generously to eliminate weed competition, and insecticides and rodenticides applied from the air protect the crops from insects and rodents. Each of these practices individually militates against the welfare of quail. Collectively they have all but exterminated the species from the better farmlands of California.

Wholesale mechanization of California agriculture is a natural concomitant of the shift from individually owned farms to massive agro-businesses. In 1970, the University of California Agricultural Extension Service estimated that 3.7 million acres of farmland in the state was owned by 45 corporations. In 1972, the census report on farm-rich Fresno County revealed that, although the county had 7,539 farms, 170 of them accounted for 65 percent of the farmland. Fewer than 3,000 farms in the entire state constitute 70 percent of California cropland. Corporate operation of large holdings generally does not foster the kind of personal husbandry required to maintain quail coverts.

Marginal farmlands, totaling some 4 million acres, have been somewhat less affected by agricultural mechanization. In the foothills and narrow valleys, there remain many small farming units with scattered fields of grain or fruit crops intermixed with grazing lands, chaparral, or forests. It is on this class of land that many of the best remaining quail habitats are found. Fields of wheat, barley, or other grains supply an important source of food in the form of crop residues. Weeds and brush bordering the croplands and protected from grazing by bordering fences add to the habitat value. The marginal farm remains a stronghold of quail occupancy.

By far the largest acreage of quail country is the 22 million acres of "grazing land," which includes a wide spectrum of vegetation types unsuitable for tillage or for silvicultural production. In this category would fall the blocks of chaparral, woodland-grass, arid grassland, and Great

Basin shrublands. The most pervasive influence on this great area has been persistent, unrelenting grazing by livestock. The adverse effects of overgrazing on quail are not as obvious as clean cultivation, but they are just as real. On casual observation, many areas of grazing land would appear to be fine quail country. Indeed, some local areas retain enough ground cover to meet quail needs. But a discouraging portion of the more arid foothill rangeland is largely devoid of quail occupancy or sustains only low populations. The situation was recognized by Sumner, who stated in 1935 (p. 290): "The life history study carried on in the Santa Cruz Mountain region indicated that one of the most important causes of quail decrease on many range lands is overgrazing by live stock."

There are many localities in the inner coastal ranges that have been fine quail range in the recent past but are now virtually without quail because of intensified grazing. Often this result follows the development of water sources for livestock in areas formerly supplied only with a few seep springs that furnished quail with drinking water but were inadequate for cattle. As an example, the Temblor Mountains south of Cholame sustained a good quail population until a pipe line brought stock water to the crest of the range. The development was justified, in part, as a source of water for wildlife as well as livestock. But the result has been severe overgrazing of the vegetation near the water troughs and nearly complete elimination of quail and chukar partridges from that portion of the range.

QUAIL IN THE CHAPARRAL

Chaparral is one of the extensive vegetational formations of the coastal ranges of California that has long been recognized as suitable cover for quail. There are, in fact, many distinctive local types of chaparral, made up of quite different species of shrubs. Along the humid coast, some of the dominant shrub genera are coyote bush *(Baccharis),* wild lilac *(Ceanothus),* poison oak *(Rhus),* and various oaks *(Quercus).* In more arid situations, these genera may be replaced by chamise *(Adenostoma),* manzanita *(Arctostaphylos),* or sage *(Artemisia).* All types of chaparral, however, are similar in having a dense, closed canopy with sparse herbaceous ground cover.

Chaparral often occurs on steep slopes or on gravel outwash fans where the raw, inorganic soils are too weak to sustain woodland or grassland communities. On these unstable soils, California's Mediterranean climate, with its long, dry, almost rainless summers, has stimulated the evolution of chaparral as the most appropriate type of protective plant cover. The scrub oak, a typical plant of the chaparral, was known to the Spanish settlers as *chaparro,* and from this term "chaparral" has come to represent dense shrubland growth.

California's dry summers are also favorable for fires, which over the centuries have periodically swept over the state's brushlands, causing all chaparral species to become adapted to fire in one way or another. Destruction of chaparral by fire is normally followed by slow recovery, which works differently in the different species. Some chaparral shrubs crown-sprout after burning, whereas others regenerate by seeding.

Chaparral has always been an important component of the quail habitat in the coastal ranges of California, and to a limited extent in the Sierra foothills and mountains of northern Baja California (see Fig. 18). The impact of land use patterns on the chaparral type has had a profound effect on quail welfare.

As regards fire in the chaparral, there are highly diverse opinions as to the frequency and extent of burns in aboriginal times. Willis Jepson (1930), the dean of California botanists, designated chaparral as a fire type of vegetation that evolved in the semi-arid California foothills. He notes (p. 114):

The vast chaparral association in California and Lower California, in its present life-form, is a product of Pleistocene and Recent periods. The characteristic struc-

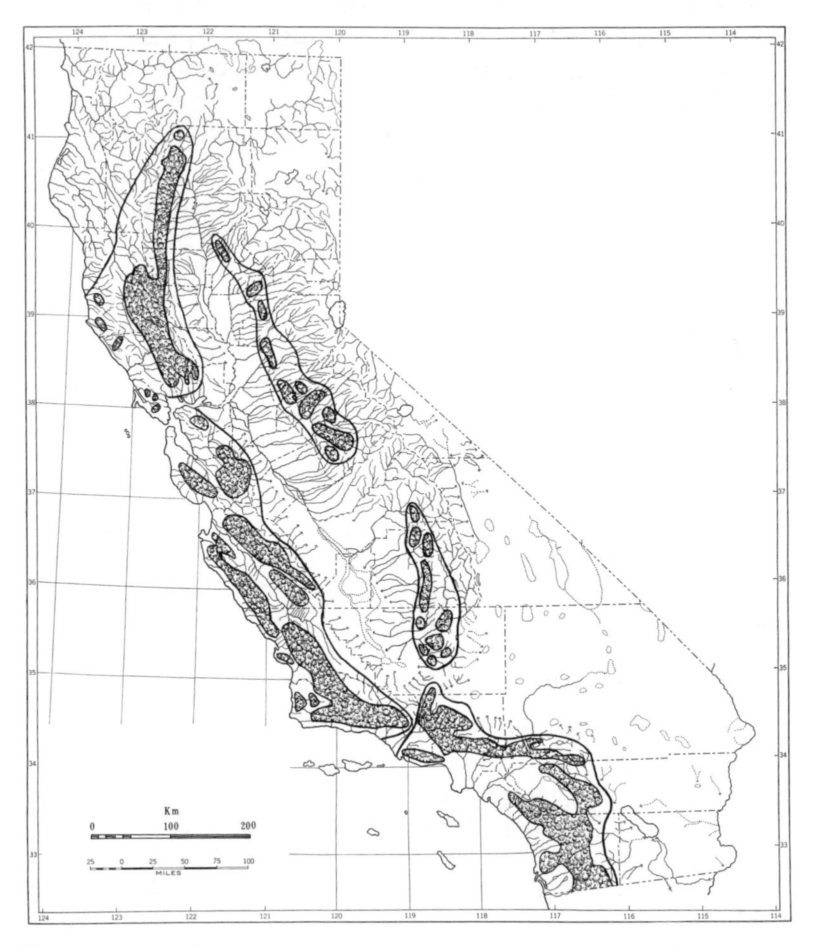

Figure 18. Map of the principal chaparral areas in the range of the California Quail. (After Sampson, 1944.)

tures appear to have developed in association with a progressively drying climate—decreased precipitation and decreased humidity. . . . Concurrently with these phenomena wild fires (originated by lightning or by the native tribes during the period of human occupation) played an important role in the development of the chaparral.

Some anthropologists and historians, including Kroeber (1925), Stewart (1941 and 1963), and Bolton (1927), feel that burning by aborigines was widespread in California. Lewis (1973) develops the theme of aboriginal use of fire as a sophisticated tool of vegetation management by certain tribes.

Cooper (1922) made the first comprehensive study of the chaparral, reported in his classic paper *The Broad-Sclerophyl Vegetation of California*. He makes a strong case (p. 74) for the existence of chaparral as a true climax form of vegetation within its characteristic climatic zone, but brings out the interesting point that frequent fires may have brought about an extension of chaparral along the mesic fringe of its range, bordering forest zones, while conversely the frequent burning in more xeric situations may have led to replacement of chaparral by grassland. He states (p. 80):

By far the most important cause of destruction [of chaparral] has been fire. It has been stated that fire favors the extension of the chaparral at the expense of the forest. It is also true that fire, if it occurs with great frequency, favors grassland at the expense of the chaparral. A single burning of chaparral will result merely in a crop of stump sprouts and greater density than before, but yearly burning will inevitably destroy the brush completely and prevent invasion by it. Cattlemen and sheepmen in the early days, according to unpublished Forest Service reports, were accustomed to fire the brush annually in the foothills to destroy it and thereby improve the grazing conditions. This resulted in a great increase of grassland at the expense of the chaparral. Such recent events, however, are of small importance compared with the effects produced by the aboriginal population. The following quotation from Jepson (1910) is of interest in this connection: "The herbaceous vegetation [in the Great Valley] in aboriginal days grew with the utmost rankness, so rank as to excite the wonderment of the first whites. . . . This dense growth was usually burned each year by the native tribes, making a quick hot fire sufficiently destructive to kill seedlings, although doing little injury to established or even quite young trees." . . . Here we have suggested the cause of destruction of the chaparral, or the prevention of its establishment.

In contrast to these viewpoints, however, are the opinions of other authorities who acknowledge that some Indian burning did occur but minimize the extent and impact of the practice. Sampson (1944:18) states: "Critical review of the mass of documents published from 1542 to about 1853 leads to the conclusion that California Indians burned vegetation, limitedly at least, to facilitate hunting, to secure native plant foods, and to clear small areas of woody vegetation for the growing of tobacco. But these same documents indicate that the fires were seldom extensive." Shantz

Figure 19. Fuel break through chaparral near Julian, San Diego County. Quail feed along the edges of this clearing. R. Wakimoto photo.

(1947) elaborates the record of aboriginal burning and concludes, in agreement with Sampson, that many of the fires were small and were set to manage vegetation on specific local areas for higher food yield or to attract deer for easier hunting. Burcham (1959) further documents the known history of aboriginal burning in California and presents two interesting maps showing those areas where the Indians were known to burn and other areas where apparently fire was not customarily used. He says (p. 182): "Contrary to widely prevalent belief, many California Indians did not employ fire for driving game when hunting. The communal drive with fire was entirely absent from the central coast region of California; nor was fire used by some tribes of the upper Sacramento Valley. Among tribes who did use fire to drive game, some employed it only when hunting rabbits, squirrels, and other small animals among tules or in grassland; others used it only when hunting specific kinds of big game. In any event, use of fire in hunting was circumscribed by a great amount of ritual which tended to limit its application; and no evidence has yet been found to indicate its widespread use for hunting in brush or forested lands.''

In another paragraph, Burcham concludes: ''When the evidence is carefully reviewed only one conclusion appears reasonable: Fire is but one factor of the environment; trees and other vegetation persist in spite of the fire, rather than because of it.''

Whatever the early fire history may have been, all authorities agree that fire severity increased in the post-settlement era of California history. During and following the Gold Rush, the slashing of timber and accelerated careless dissemination of wildfire threatened to consume the woody vegetation of California, and much of the stored fertility of the ages was lost in the ashes. Burcham (1959:181) elaborates:

> Great areas were burned during these years, especially in chaparral and forests, by many people and for many different reasons. . . . This extensive burning, together with disturbances due to mining, grazing, lumbering, and settlement, had far-reaching effects on the natural plant cover of large tracts of land. Particularly in the foothills and coniferous forests, appearance of the country was so altered that in later times the original character and condition of the natural vegetation have been misunderstood quite generally.

When finally the conservation era was born after the turn of the century, one of the first major tasks was to gain some control over destructive wildfires. Forest protective practices were extended and perfected through the early decades of the 20th century, and by the 1940's an efficient and elaborate fire control system was in operation, supported cooperatively by federal, state, and private agencies. The effect was a substantial reduction in the number of wildfires that escaped suppression, but unfortunately those that did tended to be large and highly destructive because of fuel accumulation during long periods of non-burning (Rothermel and Philpot, 1973), coupled with the invasion of highly flammable annual grasses.

Quail habitat in the chaparral suffered in two ways. The intrusion of the annual grasses and the thickening of unburned chaparral itself tend to suppress forbs, including the species of principal value in supplying quail foods. Muller (1966) and Muller and Chou (1972) have shown that dense mature chaparral, particularly of chamise and manzanita, may have a chemically inhibitory effect on the growth of understory ground plants. But even more damaging to quail welfare were the hot and often extensive wildfires that escaped suppression and that swept clean all quail cover over enormous areas. Seed-producing forbs invaded the burns (Vogl and Schorr, 1972), but these were of scant value to quail in the absence of cover.

In the 1950's, there began to emerge a new philosophy of chaparral management which emphasized conversion of the native brushlands to grassland for the purpose of improving livestock forage. When brush conversion is applied to large blocks of land, the California Quail is essentially eliminated.

Figure 20. Chamise chaparral stripped of quail cover by a wildfire, Upper Carmel Valley, Monterey County. R. Gutierrez photo.

Taken all in all, the changing patterns of land use in the chaparral type have not been advantageous to the quail. Non-burning followed by excessive burning and brush conversion, when applied to large blocks of chaparral, all depress carrying capacity for quail in one way or another. Superimposed on these different philosophies of treating the woody chaparral species has been the persistent pressure of livestock grazing (frequently overgrazing) of the herbaceous ground cover growing in or adjoining the brushlands.

THE INVASION OF BROME GRASS IN WESTERN RANGELANDS

A primary factor that intensifies the damage of wildfire in arid ranges is the intrusion of brome grasses *(Bromus tectorum* or *B. rubens)* and other alien grasses in the zone of rainfall below 10 inches per annum. It has been mentioned previously that bromes can outcompete the forbs and broad-leaved weeds that are so essential in producing quail food. Additionally the new grasses, particularly the bromes, brought adversity to the quail in the form of flash fires. The bromes, which mature early and are dry throughout the hot California summers, are probably the most inflammable of any range growth. Highly competitive, they thrive in woodlands and shrublands, commonly producing dense stands in close proximity to whatever trees or shrubs constitute the established woody growth. Well-adapted to intensive grazing, the alien grasses have invaded practically all types of semi-arid rangelands, including forest, woodland, chaparral, grassland, shrub-grassland, and desert-shrub. They have become firmly established as the most common type of annual growth throughout the more arid ranges of the California Quail. A main effect of this grass invasion has been to produce an additional supply of fuel for a new type of range and forest fire never known in the California Quail country prior to the coming of the white man with his exotic plants.

In the summer of 1966, a grass fire swept over some 15,000 acres of quail range in northwestern Kern County. All but a few spots of the shrub growth, particularly the sage, juniper, and atriplex that made this superb quail country, were killed. In 1974, after eight years, there is no sign of new junipers. New stands of atriplex have barely started, and the California sage or artemisia is still no more than about half its former growth. Only meager remnants of the pre-fire quail population remains.

In the summer of 1968, a comparable wildfire cleared off another highly productive piece of quail range near Shandon, San Luis Obispo County, which included areas of chaparral and pine-oak woodland with patches of California sage. After six years, neither the brush nor the quail have returned.

Along with the plow, cow, and sheep, wildfires must be accorded due prominence in the general removal of protective shrub cover that has contibuted to the decline or disappearance of the California Quail on the more arid sections of its range.

LOGGING IN FORESTED AREAS

Unbroken forest, like dense chaparral, is often deficient in both food and ground cover for quail. In pre-settlement days, quail were scarce or locally absent in the humid forest zone of northwestern California and southwestern Oregon. Merriam (1967:168) interviewed old Indians living in Humboldt County and was told that the California Quail was a "newcomer in their country. They have no ancient name for it." Elsewhere (p. 172) he states: "The valley quail did not inhabit the redwood forests of the Lower Klamath country but came in after the whites had made clearings."

Logging has the effect of permitting forbs and ground cover to grow, thereby supplying quail habitat where none existed before. The California Quail is not typically a bird of heavily forested lands, but it does occur along forest borders and can take advantage of favorable openings created in the timber stands. In the foothills of the west slope of the Sierra and in some parts of the coastal ranges, quail have profited by logging operations that permit the growth of legumes and various weeds in the openings. These food plants are stimulated partly by the removal of the overhead canopy and partly by the disturbance of the soil surface in the process of logging.

Generally speaking, it can be said that, at elevations below 3,000 feet, silvicultural practice that results in creating openings in the forest may be of positive benefit to quail, particularly if the slash is piled into brushpiles to create secure ground cover where none existed before.

TRENDS IN LAND USE

Considering the range of the California Quail as a whole, and the patterns of land use that are developing, the quail habitat is still deteriorating. Advantages that derive from forest practices in some areas and ranch establishment in others are more than offset by the continuing depletion of habitat by intensive agriculture, grazing, unregulated wildfire, and brush clearing. The basic problems of habitat deterioration that were recognized by Sumner in 1935 and by Emlen and Glading in 1945 have continued unabated over most parts of California and Baja California.

It is unrealistic to think that fortuitous land use practices, dictated primarily by economic return, are going to change for the better in the future. Quail habitat will be maintained only where a deliberate and planned effort is made to do so. However, within the context of sound land use, the habitat needs of California Quail can be provided. If motivated to do so, a landowner or a land operating agency can restore and sustain quail range with very little cost or loss in economic return. But quail management, to be effective, requires continuous, conscious husbandry and attention.

PART II
NATURAL HISTORY

5
SOCIALITY IN CALIFORNIA QUAIL

GREGARIOUS TENDENCY IN THE SPECIES

Except during the nesting cycle, the California Quail is a highly social and gregarious bird. In fall and winter, when the birds are grouped into coveys, the behavior is highly organized with social compatibility the characteristic feature. This behavior changes profoundly with the onset of pairing in early spring. In March, when the pairs are becoming well established and males are guarding their mates, aggressiveness among the males becomes the characteristic behavior, and it prevails in some degree until termination of the breeding cycle. When the young emerge, the parents plus chicks constitute a firm social grouping. As summer gives way to autumn, family groups combine to form coveys, which remain the social entities until the time of reproduction recurs. When quail are flushed and scattered, their main preoccupation following the disturbance is to reassemble with their companions.

In discussing flocking behavior in birds, Emlen (1952:162) envisions the individual as being subjected to social "forces" that control the formation and size of flocks. He speaks specifically of "a positive force of mutual attraction and a negative force of mutual repulsion, interacting in the formation of bird flocks." He goes on to illustrate how the flock, held together by strong mutual attraction, can be caused to disintegrate when individual birds are changed physiologically so as to weaken the "positive force" of mutual attraction. Specifically, Emlen and Lorenz (1942) implanted pellets of sex hormones under the skin of California Quail in normal winter coveys, and the treated birds of both sexes became pugnacious and aggressive, seeking mates and actively rejecting the company of the covey. Thus the gregariousness so clearly evidenced in winter quail coveys is an evanescent characteristic, subject to quick and substantial change under seasonal and hormonal influence.

ADVANTAGES IN SOCIAL ORGANIZATION

The gregarious habits of the California quail, and for that matter of other American quail species, lead to the assumption that the birds gain some definite advantages from covey formation during non-breeding seasons of the year. Sumner (1935) paid particular attention to this facet of quail life. He emphasized the tendency for quail to imitate one another (p. 206):

> The reactions of each member of a covey are also indicative of a fairly well disciplined social unit. For example, with the exception of a few inexperienced young of the year, when one quail gives the warning *Ku-rr!* note and "freezes," they all freeze, and if one bird bursts into the air and sails away from an enemy the others usually follow. That this disciplined obedience has its beginning in the strong tendency of young birds to imitate their parents, which in turn are more than usually furtive and wary during the very time that the young are growing up, has been pointed out previously. The fact that late winter coveys tend to be more alert and disciplined in their reactions to danger is explainable partly on the ground of gradual learning by the young, and also partly from the fact that the more sluggish and least cooperative tend to be weeded out by enemies.
>
> Birds which have become separated from their fellows, and individuals which have been liberated alone in new localities, show marked uneasiness and run about distractedly, giving the assembly call in loud, far-carrying tones. The anxiety of a flock to reassemble at roosting time is so great that it will often do so even in the face of a danger which still threatens.
>
> The imitative tendency asserts itself throughout life, as is shown by the manner in which the adult birds will follow one another as they move from place to place when feeding. . . .

Sumner discusses the question of "leadership" in covey organization and concludes that the leader of the covey at any given moment is a matter of chance. If a bird of either sex moves off in a determined manner, the whole covey may follow. Several "leaders" may take off at once, splitting the covey temporarily or even permanently, but if a pioneering bird is not followed by others it abandons its exploring and hurries back to rejoin the covey.

Sumner goes on to note that some of the social habits of quail are decidedly cooperative. Thus when one bird finds an unusually good supply of food it often calls the others to it, in the manner of a domestic hen calling her chicks to a newly discovered food source. Likewise, a member of the covey that perceives danger will warn the group by appropriate vocalization. Sumner recounts an incident of a covey that was flushed into the canopy of some live oak trees and kept there for over an hour by the repeated, low-

Figure 21. A covey of California Quail clustered in the shelter of a small brushpile.

pitched warning calls of one bird that could see him (Sumner) waiting quietly below. All in all, the covey unit must hold substantial social advantages for the members comprising it.

SENTRY DUTY OF COCKS

One of the striking social habits of the California Quail is the sentry duty performed by cocks throughout the year. When a pair withdraws from the covey and goes about the business of nesting, the cock becomes increasingly watchful for any possible threat to its mate, spending much time standing motionless on the alert while the hen feeds or constructs her nest. If all is well he utters occasional soft clucks of reassurance, but if a hawk or other danger appears the cock emits warning *pit-pit* notes and the hen dives for cover. This solicitous activity intensifies during the nesting period to the point where the cock spends relatively little time feeding and much time on the alert. During incubation, the cock waits patiently for the hen and accompanies her when she takes a feeding break; he does sentry duty while she feeds. What nourishment he gets is gathered at odd times during the day or just before going to roost in the evening.

Sumner (1935:209) states:

At hatching time the duties of the male as guardian are enormously increased, for he must now maintain a constant vigilance in behalf of the young birds while they forage throughout the long summer day. Food is eaten by him only intermittently, and he loses weight accordingly. At the least sign of alarm he gives the sharp, warning *kur-r!* note which sends the family scurrying for shelter; and when the coast seems clear, it is he who unobtrusively creeps out of cover before the others and remains motionless for many minutes, on the lookout for any return of danger. If he commences again to utter at intervals his soft clucking note, the others venture out; otherwise they do not.

When the young have developed to nearly adult size and have themselves learned to be somewhat alert to danger, the intense preoccupation of the cock with guardianship wanes somewhat. But it is still he who is most likely to perceive danger and to issue a warning.

As family groups form into winter coveys, the sentry duty is assumed by one or another of the adult cocks in the covey. Frequently when a covey is loafing in dense cover during mid-day, a cock will be seen perched conspicuously on a vantage point where he can observe the approach of a predator. Some quail hunters carry binoculars to help locate the sentry cock and thereby locate the covey. Rarely if ever have I seen more than one cock on duty at a time, but the manner in which the sentry is chosen or volunteers is unknown.

Richard Genelly tells me that near his summer cottage on the Trinity River he frequently sees *hen* quail serving sentry duty for the covey. This I have never observed.

CALLS OF THE CALIFORNIA QUAIL

Sumner (1935) described 10 quail calls and explained the context in which each call is given. Williams (1969) amplified and elaborated the vocal repertoire of the quail, and by means of audiospectrographs he differentiated 14 calls, defining each in terms of causation and function. The serious student of avian communication is referred to Williams' excellent report. For our purpose here, I briefly describe 11 of the most common calls of the California Quail, following Williams' terminology.

1. *Aggregation and contact calls*

Cu-ca-cow (assembly call).—This loudest and best-known call of the California Quail is given by both sexes throughout the year. Birds in covey frequently call *cu-ca-cow*, especially in early morning and late evening. Hunters often simulate the call, listening for a response to locate a covey. Scattered birds use this call to reassemble the covey. During the mating season, an unattached female will announce her availability with a low *cu-ca-cow* in response to the male *cow* call.

Ut-ut (conversational note).—The *ut-ut* is a low-frequency, repetitive call used by both sexes to maintain contact. It is frequently heard when a covey is feeding through weedy cover that inhibits visual contact. An incubating female leaving the nest uses this call to attract her waiting mate.

Tu-tu (food call).—A bird of either sex may give this call when a good source of food is discovered. Others are attracted to share the food. Parents call their chicks to food with the *tu-tu* signal.

2. *Alarm calls*

Pit-pit (alarm note).—Next to the *cu-ca-cow*, the *pit-pit* call is the most frequently heard. It consists of a series of metallic-sounding *pits*. Appearance of a cat, a dog, or a hunter will often initiate *pit-pit* calling.

Kurr (signifying extreme fright).—This call is most frequently elicited by the sudden appearance of a hawk. Williams (p. 638) refers to it as an "aerial alarm call." Sumner (p. 204) comments that "when the field worker hears this call in the Santa Cruz Mountains he will be correct about 75 per cent of the time in concluding that a Cooper's or Sharp-shinned Hawk is after the birds."

Pseu-pseu (distress cry).—Both sexes held in the hand or seized by a predator may give this loud call of distress.

Put-put (extended warning).—This call, which sounds like a distant outboard motor, is used to keep scattered birds hidden and motionless when danger continues.

3. *Agonistic and sexual calls.*

Cow call (male mating call).—The *cow* call is given by unmated males seeking mates. It is similar to the last note of the *cu-ca-cow* series, but louder and longer. In years of high reproduction, *cow* calls are heard persistently through the mating season, given always by bachelor cocks. Mated males do not ordinarily issue *cow* calls.

Sneeze call ("squill" of hostility).—A series of short, vehement sneezes is given by aggressive males during the mating period when dominance relationships are being established. A mated male will emit the *sneeze* call in defense of his mate from another male. Sumner (1935) calls this the squill call.

Wip-wip call (male aggression).—Another call frequently used by aggressive males during breeding season encounters is the *wip-wip*. This call is also emitted by males after copulation.

4. *Parental calls.*

Mo-mo-mo (call to scattered chicks).—This call, much like the *ut-ut,* but uttered more softly, is given by either parent to maintain assembly in the brood. The warning *pit-pit* call is used to induce scattering and hiding when the brood is threatened, and reassembly is achieved with soft *cu-ca-cow* or *mo-mo* calls.

These calls constitute the principal form of social communication in the life of the California Quail.

COVEY FORMATION

Much the best account of the process of amalgamation of broods into coveys is that of Emlen (1939:128), which I quote at some length:

> The development of broods in sub-coveys by the incorporation of unsuccessful local adults and of other broods was regulated in part by the social relations and

Figure 22. Male California Quail giving the "cow" call from an elevated perch. This bird is seeking a mate. M. Erwin photo.

tolerances of the birds during this period. Immediately after hatching a definite social barrier was thrown around each brood by its parents. This was gradually lowered as the chicks grew older.

Among the various classes of local adults, the first to be admitted through the circum-brood barriers were the resident pairs that had been unsuccessful in nesting; indeed, it is questionable whether these birds were ever rigidly excluded. Such broodless adults, particularly the males, showed great interest in young chicks and, when admitted, made as good guardians as the true parents. . . .

The incorporation of miscellaneous adults into the sub-coveys formed the primary step in covey formation; the second step, operating simultaneously, consisted in the combining and merging of neighboring broods and sub-coveys. Parents (and other adult escorts) were wary of neighboring broods until their own chicks were two weeks or more of age. When 3 to 4 weeks old the parents occasionally allowed their charges to mingle with others of similar age for short periods. . . .

With the condensing of the population and the evacuation of summer range, the distribution of the local quail became discontinuous, breaking up into a number of circumscribed and isolated territories. . . .

Despite this exclusiveness of sub-coveys, two mergers took place in the latter part of November under the pressure of (1), hunting on adjacent lands; and (2), drastic reduction of roosting cover through the falling of the leaves in the orchards. This finally condensed the Farm population onto four covey territories, almost identical in location and boundaries wih those of the previous winter.

To sum up, when a nesting pair hatches a brood of chicks, the family constitutes a compact and identifiable social unit. Family groups tend to keep to themselves and to avoid the company of other quail, insofar as that is feasible. Gradually, however, the intolerance between units wanes, and there is a growing tendency for families to combine in loose aggregations. Sumner (1935) says this happened to his study birds near Santa Cruz in late July, when the young were half grown. In desert situations where water is scarce, families meet daily at the drinking hole, and the process of mutual acceptance is probably accelerated.

COVEY INTEGRITY

From the late summer aggregations of family groups are formed the autumn coveys, whose identity with a given range becomes firmer as winter approaches. Experienced quail hunters know from year to year that coveys will be found in the same general localities and that other areas often appearing superficially just as desirable will rarely or never support coveys. There undoubtedly is a carry-over of individual birds from one year to the next, and they would know the terrain and tend to perpetuate tradition of covey usage and behavior. But the real criteron determining both the location and roughly the size of a covey is the nature of the range itself. Sumner (1935:212) states: "In reality the numbers of a covey and its location are governed largely by the available food and shelter rather than by the fortuitous presence of its individual members; and in this respect the covey is analogous to a river, which constantly maintains itself although the individual drops of which it is composed continually pass onward."

In point of fact, although coveys are associated with specific ranges, they often meet along borderlines, and the nature of the social contact will vary from friendly intermixing to determined group repulsion. Sumner offers data showing rather extensive interchange of members between five coveys living on his intensive study area near Santa Cruz. Howard and Emlen (1942), on the other hand, conducted elaborate observations and experiments on intercovey social relationships among six coveys living on or near the Davis campus of the University, and they demonstrated a high degree of specificity in covey composition and intolerance between the groups. Birds lost from one covey would be attracted to follow any other covey that was encountered, but for some weeks they would be recognized as aliens and forcibly excluded from intimate association with the group. The intolerance gradually diminished, and after four weeks or so the alien would become an accepted covey member. Attacks on aliens were always

Figure 23. A covey of about 160 quail and a few mourning doves flushing from an atriplex patch.

mounted by birds of the same sex as the alien. A bird acclimated to a new covey, upon returning to its original covey, would be recognized and accepted after an absence of a week or so; but if the absence had been several weeks the individual was treated as an alien.

The integrity of small covey units which Howard and Emlen observed in the vicinity of Davis is not necessarily typical of the species in all situations. At Shandon, where winter populations tend to be large and more loosely structured, there is little or no evidence of hostility between birds of different coveys. On the McMillan ranch, where adjoining coveys vary in size from 25 to over 1,000, the numbers of the respective flocks fluctuate widely from time to time, indicating that interchange is common. The number of birds counted at a particular location may vary from day to day, with high counts suggesting a banding together of smaller groups, while low counts indicate dispersal.

In January of 1973, after a year of extreme drought and population decline, all the quail found in a mile-long stretch of quail habitat on the McMillan ranch were concentrated in a main central covey, where roosting cover was particularly favorable and where supplemental food was provided each evening. Concurrently, quail were absent from a neighboring area where evidence of predation was notably high, and where a covey had regularly spent the winter for more than two decades. In the two years following this crisis, 1973 and 1974, rainfall was above average and the quail populations increased markedly. By late fall of 1974, at least four coveys were occupying this same mile-long area, making a total population at least double that of the 1972–73 low.

This year-to-year fluctuation in numbers and size of coveys, with correlated restriction and expansion of occupied winter range, has been found to apply generally to quail populations inhabiting comparable areas throughout central and southern California.

COVEY TERRITORIALITY

Whereas a covey of California Quail is a social unit with a preferred home range, there is no evidence that the occupied area is defended from the incursion of other quail. Rather, it would appear that where intercovey hostility does occur it is the covey itself which is defended, in the sense that strangers are not made welcome to the group. In the classic sense, "territoriality" is the defense of an area. California Quail do not seem to show this proprietary feeling toward their home ground.

Occasionally two coveys meeting along the boundary of their respective home ranges engage in a general melee. Genelly (1955:276) noted such an encounter on October 26, 1952, along a road that bounded the ranges of two coveys. Emlen (1939:129) describes two covey encounters observed in his intensive study of the quail population on the Davis Campus. In both cases (on September 15 and 16), incipient coveys attacked neighboring groups of parents and young in a lively and spirited manner. These autumn encounters seem to relate to the period of covey formation and home range establishment, but Emlen felt that the motivating factor was a form of "social exclusiveness," not defense of an area per se. In winter, after coveys were well established, he observed covey encounters in which the groups of birds were mutually repelled, but there was no evidence of general aggressiveness in defense of a geographic territory.

COVEY SIZE

As we have just explained, the number of birds comprising a covey of California Quail is a shifting and variable entity. Coveys are usually larger in autumn than they are in late winter, especially in areas subjected to hunting. They are decidedly larger in years of good hatch than in poor years. Likewise, a covey may be scattered and split by hunting or by predators, and counts made one day may not correspond to those made the next. Despite some evidence of mutual exclusiveness of coveys, there are many instances of two or more coveys combining into large aggregations. As a rule, however, covey counts made throughout the fall and winter under a variety of circumstances and in different areas are remarkably uniform.

In arid regions where drinking water is scarce, all the quail in a considerable area may be forced to assemble at a local water source for the daily drink. Following a wet year when large numbers of young are produced, the resultant traffic is heavy, and several hundred birds may be thrown together by circumstances beyond their control. Social friction between family groups is eroded away, and by autumn a functional covey of two or three hundred birds may form. Nowadays, aggregations of this size occur

Figure 24. Ian McMillan's "home covey" of approximately 1,000 birds in February, 1972.

only in favorable years in areas that have retained a high carrying capacity. In the pioneer era described in Chapter 2, this phenomenon apparently was common.

Under intensive management and protection, abnormally large coveys may still be produced. On the Ian McMillan ranch near Shandon, a home covey has been developed which numbers from 800 to 1400 birds each winter. Several acres of cover plantings around the homestead serve as the home base for this mob. The birds receive approximately 30 pounds of barley each evening in winter as a supplement to the natural food supply. This situation will be elaborated elsewhere, but it is mentioned here to illustrate that the "typical" covey may be enormously increased by artificial management.

In most situations, however, under extensive management or no management at all, the species will tend to occur in coveys of from 30 to 70 birds, averaging about 50. Table 2 summarizes some sample counts made during this study or recorded in the literature. The average wild covey of California Quail is much larger than the average covey of Bobwhites. Rosene (1969) tallied 2,815 coveys of Bobwhites in the southeastern United States and found the average size to be 14.3 birds.

SEASONAL MOVEMENTS OF QUAIL

In arid ranges, quail coveys tend to form in late summer in the vicinity of drinking sites. Many broods dependent on a given water source assemble there daily, and, as the chicks begin to mature, the social barriers between family units disappear, and one or more coherent coveys will be formed. As long as the birds are dependent on water, the coveys remain localized, which makes them vulnerable to predation. If the hunting season opens

TABLE 2.
Covey size in California Quail

Source	*Area*	*Year*	*Number of coveys*	*Average size*
Grinnell et al. (1918)	California	1912–18	?	29
Sumner (1935)	Santa Cruz Co.	1931–35	4	34.8
Emlen (1939)	Davis campus, Yolo Co.	1937	4	28.2
DeLong (pers. comm.)	Monterey Co.	1971–73	16	40
Leopold	Penobscot Ranch, Georgetown, Eldorado Co.	1968–72	43	26.4
Glading (1938a)	San Joaquin Exp. Range, Madera Co.	1937	?	30
Erwin (Appendix C)	Camatta Ranch, San Luis Obispo Co.	1973	19	66.5
McMillan	Shandon area, San Luis Obispo Co.	1970–74	50	72.8*

*Excluding the subsidized "home covey," which varies in size from 800 to 1400 each winter.

before the fall rains begin, such coveys are also highly vulnerable to shooting, since they are concentrated and have no real option to move in adjustment to hunting pressure. With the beginning of the rains, however, water dependence disappears, mainly because new drinking places are formed or the birds can satisfy their water needs by eating succulent new greens. Then, coveys often make substantial moves to areas of suitable cover and food that were unusable in the dry season for lack of water.

Seasonal movements of quail are sometimes substantial. Savage (1974) banded 1,889 California Quail in Modoc County, California, between 1966 and 1972, and he had returns as far as 11 miles from the banding site. Considerable movement within 5 miles of the trap location was established by 37 observations and 20 band returns. In various quail-banding projects pursued by the California Department of Fish and Game in central California, movements of up to 5 miles were recorded. Most long movements of individual birds probably occur when the coveys disperse for nesting.

Another seasonal phenomenon that initiates substantial changes in quail location is the arrival of the fall migration of two species of hawks that commonly hunt quail—Cooper Hawks and, to a lesser degree, Sharp-shinned Hawks. Although common on most quail ranges in fall and winter, these species breed largely in wooded country and are rare or absent in summer from the more open quail ranges. In three seasons of extensive observations of quail nesting areas in the Shandon region, Erwin noted only one nesting pair of Cooper Hawks, and this was in woodland habitat. The few nests of this species that have been recorded in the Shandon region have all been in either woodland or riparian habitat far removed from the more open types of quail country.

In summer and early autumn, quail are prone to forage far from dense cover in search of preferred forage supplies. However, with the arrival of the migrant Cooper and Sharp-shinned Hawks, which usually occurs in August, there is an immediate shrinkage of the available quail range, with Cooper Hawks having by far the more dramatic impact. The quail quickly localize their movements in the vicinity of secure ground cover when Cooper Hawks are present. The dispersal of nesting pairs of quail in the spring occurs after these migrant predators have moved out. In areas where the summer ranges of quail and Cooper Hawks overlap, the quail are never free to wander far from cover.

6

COVEY BREAK-UP AND THE NESTING SEASON

WEAKENING OF THE COVEY BOND

The first incipient signs of sexual awakening in California Quail usually appear in late January or February in the Shandon area. Individual males within a covey will be seen to make aggressive runs at other males, chasing them for a few feet before turning away. Concurrently, as the covey moves about, there will be seen a certain number of pairs that appear to stay together in closer proximity than other members of the group. These behavioral developments are subtle and seemingly casual for several weeks, being most noticeable on sunny days when the birds are loafing about, not engaged in serious foraging.

The time of initiation of the pairing process appears to vary from year to year and from place to place. In warm, dry spring weather pairing seems to start early. For example, at Shandon in March of 1972, chases among males were frequent, and there were numerous pairs of birds seen about the ranchyard. In contrast, in the spring of 1973 pairing was deferred because of extended cold rains. That year, however, resulted in high production of young when nesting did get started (see Appendix C).

As regards differences in time of breeding between localities, there appears to be some relationship to altitude, latitude, and local climate. Sumner (1935:213) states:

About the last week in February, if the weather has not been unusually cold, the first signs of pairing commence to be in evidence. . . . The dates given here for the various stages of the reproductive cycle are averages of three years' observation in the Santa Cruz Mountain region, at 1,700 to 2,000 feet altitude. For other regions they will be slightly earlier, or later, according to the climate. Thus in San Diego County the whole cycle occurs about three weeks earlier than in the region under discussion. In Oakland, Berkeley, and Palo Alto, California, whose respective climates are warmer and the vegetation cycles earlier than in the Santa Cruz Mountains . . . the reproductive season commences about seven to ten days earlier.

Anthony (1970a) found the reproductive cycle in eastern Washington to occur a month later than that at Berkeley.

Casual fighting between males and the loose formation of pairs continues through March but intensifies as the actual breeding season approaches. Warm weather stimulates a sharp increase in fighting as the males grow more truculent. As explained later in this chapter, the change in behavior is associated with the increased secretion of sex hormones under the stimulus of lengthening periods of daylight.

PAIRING BEHAVIOR

The process of pair-formation generally occurs without a great deal of display or ostentation on the part of the members of the pair. Most of the action occurs between males. Sumner (1935:213) describes the process as follows:

Apparently each male tends to attach himself to a particular female which he follows about, while constantly endeavoring to drive away any other male which comes near. The fighting between males occurs almost any time that two males get within two or three feet of each other. At such times one male runs at the other with lowered head and ruffled feathers, but now, instead of running away as formerly, the other will often stand his ground with lowered head, returning peck for peck amid a chorus of excited clucks and loud *squill!* notes. The birds rush at each other and dodge with great rapidity, sometimes even fighting for a moment breast to breast, when they peck at each other's faces and leap simultaneously into the air like fighting roosters. . . .

The vanquished male is not usually chased beyond ten or fifteen feet, but occasionally the pursuit lasts longer as the winner follows every twist and turn of his foe with furious determination. On one rare occasion the loser was actually pursued through the air for fifty feet before the chase was abandoned. Immediately upon the conclusion of a battle the victorious one gives the *squill!* note with great energy and defiance. . . .

During this period the covey still preserves its identity but it is now split into discordant groups and torn by the excitements accompanying the gradual development of sexual awareness, so that its final dissolution is imminent. Although the writer was not able to observe the details of the process, it is clear that as a result of several strenuous weeks of fighting and calling, each successful male wins a place by the side of a particular female and prevents all other males from approaching her. As in the case of the bobwhite, female California quail take no part in the scuffling of the males, nor do they fight among themselves, but that they are interested bystanders is indicated by the fact that they frequently give the assembly note in response to that of the males and often come running in answer to the calls of the latter.

Raitt (1960) elaborates the forms of hostile behavior between cocks during the process of mate selection. He recognizes three types of hostile expression: *Chasing*—the pursuit of one male by another, as described above; *fighting*—involving physical contact, as in a contest between fighting cocks; and *nudging*—when a cock moves between his mate and a rival and shoulders the latter away. Taken all in all, the behavioral patterns of California Quail during the mating process are simple and unsophisticated as compared with many other gallinaceous birds.

Copulation occurs throughout the period of pair formation, beginning, according to Raitt (1960:288), shortly after the preliminary pair bonds have been formed:

In no case was any pre- or post-copulatory display given. Typically, the two birds foraged close together with the female leading. The female stopped feeding, settled into a low squatting position, and remained motionless. The cock, on observing the behavior of the hen, stopped feeding, straightened up, remained motionless for a very short time, and then stepped forward and onto the back of the hen. Copulation was accomplished in a few seconds; the male stepped off and feeding was resumed immediately by both birds. In coition itself the female raised her tail into a vertical position; the male maintained his back in a nearly erect posture, bent his neck down, grasped the nape of the hen with his bill, and moved his pelvic region back and forth several times. Copu-

Figure 25. A copulating pair of California Quail. M. Erwin photo.

lation may occur when a pair is isolated from the covey, but more often it was observed while other birds were feeding nearby.

In reference to this last point, it should be noted that Michael Erwin observed copulation more frequently in isolated pairs.

In the Bobwhite Quail, Stoddard (1931:17) describes a complex display performed by the male in front of the female in advance of copulation. He interprets this action as a form of courtship. Genelly (1955:267) observed a very similar type of display of unmated California Quail cocks toward hens confined in wire cages. The plumage is fluffed out, the head is lowered, the wings are dropped to the ground, and the tail is spread above the back as the cock ''rushes'' the confined hen. These displays are observed so rarely that they can scarcely be considered a normal part of pre-copulatory behavior. Raitt feels that this performance may be an expression of intense interest toward a confined (hence unavailable) hen on the part of an unmated cock. The mass of evidence points toward a pattern of normal behavior between mates that is subdued and rather casual.

THE UNMATED COCKS

In a wild population of California Quail, cocks almost always outnumber hens. The reasons for this disparity will be discussed in Chapter 8, but for the present our concern is with the extra cocks that fail to acquire mates. When the coveys scatter and mated pairs go about the business of nesting, the unmated males announce their interest in obtaining mates by the characteristic *cow* call. Generally, this single note is heard only during the mating season. The *cow* call appears to serve as an advertisement of the availability of unmated cocks in the event that a hen for one reason or another is unattached. It is analogous to the *bobwhite* call of unmated Bobwhite males (Stoddard, 1931:98). This social mechanism assures that no female goes unattended. Glading (1938a) and Genelly (1955) noted a few instances of the *cow* call being given by mated cocks, but this is exceptional.

HORMONAL CHANGES LEADING TO REPRODUCTION

The ecologic and physiologic processes by which birds time their breeding cycles have been studied at great length and in many parts of the world. Clearly, each species of bird living successfully in a local environment must have evolved a system of temporal regulation of breeding that on the average affords the greatest probability of successfully raising young. This subject is reviewed in detail by Klaus Immelmann in his essay, ''Ecological Aspects of Periodic Reproduction'' (1971). In temperate regions the vast majority of birds breed in spring and early summer, which is the period most favorable for rearing offspring. And the built-in timing mechanism which initiates reproduction is a delicate sensitivity to photo-periodism, or change in length of day. Laboratory experiments with many bird species have demonstrated that "increasing photo-period after the winter solstice is the most important signal for gonad development. . . . [T]here is no other environmental factor in any climatic region that has gained a similar importance for the immediate control of annual cycles" (Immelmann, p. 363).

Considering the quail specifically, when the short days of winter begin to lengthen, the hormones which initiate development of the sex organs begin to flow from the pituitary gland at the base of the brain. The change in light, or perhaps the lengthened period of daily activity, effects a response in the pituitary cells, which produce and release gonadotropic hormones. The latter, carried to the gonads in the blood stream, initiate growth of the testes and ovaries, which in turn release specific sex hormones that stimulate reproductive behavior. This sequence of events can be produced artificially in mid-winter by placing quail in a light-proof cage and simulating with electric light the lengthening periods of day normally experienced in spring. We may assume, therefore, that each California Quail is endowed with such a sophisticated physiologic device to prepare it for breeding in the springtime. Immelmann refers to this as the principal "proximate control of reproduction" (p. 363).

However, the actual implementation of the breeding process may be further regulated by a number of modifying factors such as weather during the nesting period, the general health and fitness of the birds, their food supply (both quantitative and qualitative), availability of suitable nesting sites, predation on eggs and young, food for the young, and others. These Immelmann calls "ultimate controls of reproduction" (p. 363), and each in turn will be considered later in this volume. For the moment, let us examine the physical evidence of sexual development that leads to nesting and egg production in California Quail.

DEVELOPMENT OF THE GONADS

Lewin (1963) has documented most precisely the anatomical changes that occur in quail throughout the breeding period. Working in the vicinity of Berkeley, he collected and dissected birds at all stages of the reproductive cycle through three breeding seasons, 1954 to 1957.

Figure 26 summarizes Lewin's data on the growth and recession in the size of quail testes. The quiescent organs in mid-winter, measuring about 4 mm in length, begin to swell in size in late February and reach a peak of development (about 15 mm) in May. Figure 27 depicts the male urogenital system in full breeding condition. By early June the size is receding, and the testes are back to normal winter dimensions by the end of August. Note that the testes of "immature" males (approaching a year of age) and those of older males follow essentially the same growth curve. Juvenile males, produced during the cycle, have very small testes in summer, but by early autumn these have developed to essentially the size of the regressed testes of birds that have reproduced.

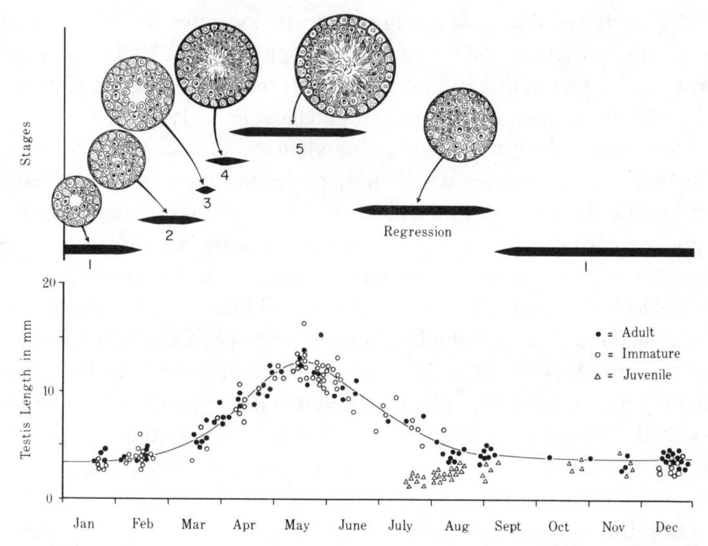

Figure 26. Annual cycle of growth and regression of the testes of California Quail. Stages of spermatogenesis are depicted above. (After Lewin, 1963.)

Also depicted in Figure 26 are cross sections of the seminiferous tubules in the testes where sperm are produced. The time when viable sperm is available for fertilization (stages 4 and 5) is approximately the three-month period of April, May, and June.

Figure 28 presents the cycle of the quail ovary, expressed in weight rather than in a linear measurement. Adult females that have gone through a previous breeding season have substantially heavier ovaries than the immature, or yearlings. But during the period of ovarian development in March and April, this difference is obliterated by rapid growth of the ovaries of younger birds, and when actual egg production begins in late April the two age classes of females are equally developed. This does not necessarily mean, however, that they are equally fecund, as will be noted later.

The period of egg laying in Lewin's sample was confined largely to the months of May and June. Presumably he drew birds entirely from a population that nested successfully in the first attempt and did not have to resort to second nesting later in the summer. We know from field experience at Shandon (see Appendix C) and in other areas that the actual period of egg production may be extended to late summer and even into early autumn, depending on weather, nutrition, and other factors. For example, Raitt and Genelly (1964) noted a deferral of egg laying near Berkeley in a year of extended cold spring rains.

Figures 29 and 30 depict the female reproductive tract in quiescent winter condition and in full breeding condition respectively.

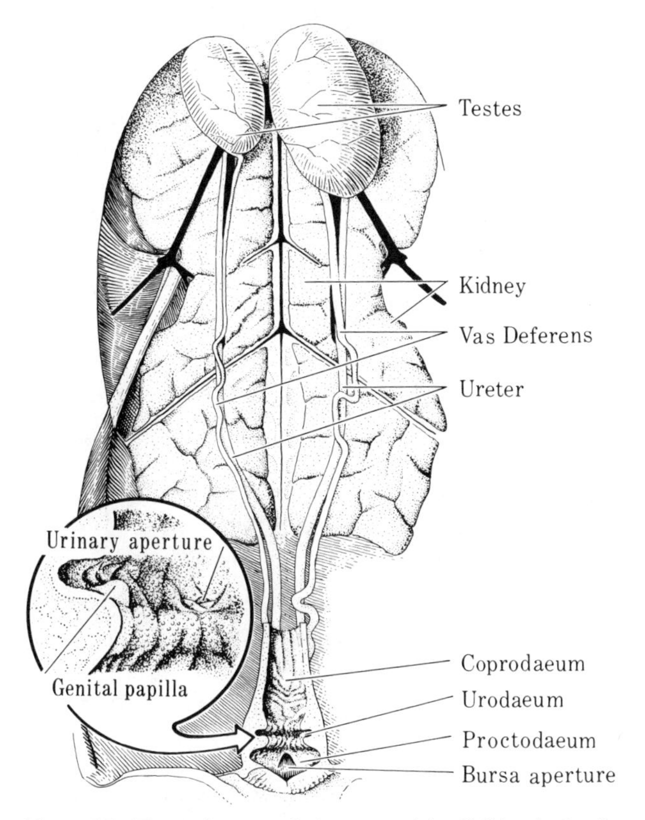

Figure 27. The male urogenital system of the California Quail. (After Lewin, 1963.)

CHANGES IN BODY WEIGHT

As would be expected, the stresses and strains of reproduction draw heavily on the energies and the body tissues of quail. Figure 31 shows the average weights of male and female quail throughout the year, plotted against a diagramatic presentation of events in the reproductive cycle.

In the case of males, there is a marked increase in weight prior to the initiation of breeding. The rise in weight of both sexes during the winter months occurs when the birds shift their diet from seeds to bulky green foods that are held for some time in the digestive tract and add substantially to the total body weight. With the initiation of pairing, there is a dramatic decrease in weight, occasioned no doubt by the preoccupation of the males with defense of the female. Fully 10 percent of the body weight is lost between March 1 and May 1. In June, body weight of males begins to build up toward the normal autumn level of 170 to 175 grams.

Females also gain weight prior to pairing and lose weight during the

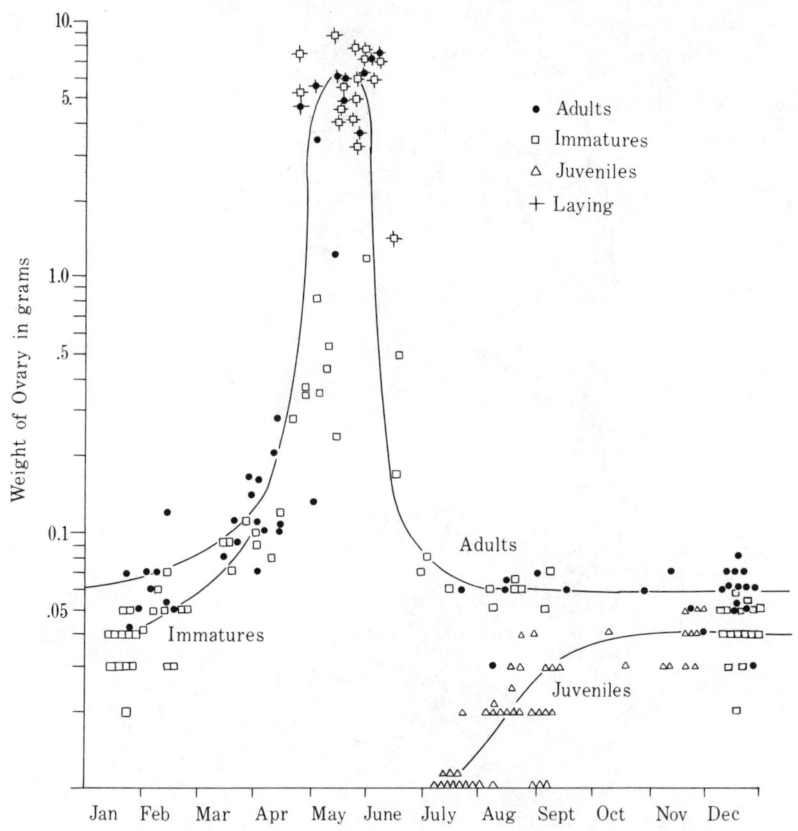

Figure 28. Annual cycle of growth and regression of the ovary of the California Quail. (After Lewin, 1963.)

tumultuous days of courtship and mating. However, when eggs begin to form in the ovary, the total weight of the female increases to its highest point of the year. The female reproductive organs add 20 grams to the weight of the bird during the peak of egg production. When laying terminates, the ovary and associated structures shrink rapidly, and the weight of the whole bird drops accordingly. Incubation is a period of continuing stress for the female, and the subsequent period of brooding and shepherding the chicks gives little opportunity for recovery of condition. It is early August, when the young are partly grown, before the hen begins to recover her normal weight, which in the Berkeley area is 160 to 165 grams. As will be elaborated later, there is substantially higher mortality of hens than cocks during the breeding season, and this may relate in considerable measure to the severity of the physiologic strain imposed on the females by the reproductive process.

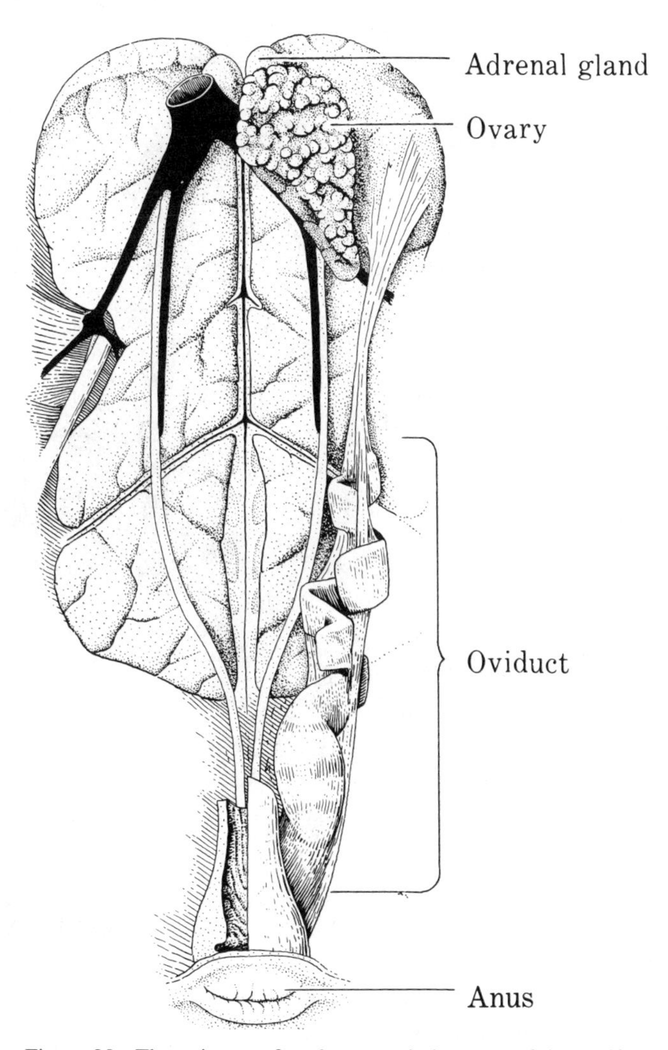

Figure 29. The quiescent female urogenital system of the California Quail. (After Lewin, 1963.)

NESTING

Grinnell et al (1918:515) describe the nest and eggs of the California Quail as follows:

Nest—Usually a mere depression in the ground, lined sparingly with grass and weed stems; occasionally a more substantially built affair, though still relatively crude, of the same materials, and placed on a log, stump, or in a brush pile; rarely in trees or other situations above ground.
Eggs—6 to 28, usually 13 to 17, pointedly oval, measuring in inches, 1.10 to 1.40

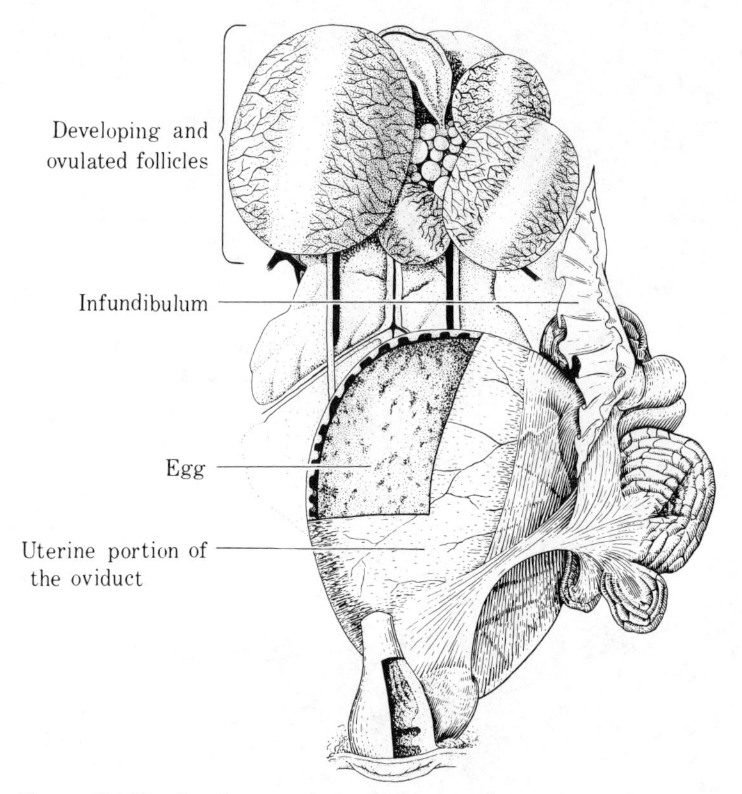

Figure 30. The female urogenital system of the California Quail during the reproductive period. (After Lewin, 1963.)

by 0.84 to 1.02 (in millimeters, 27.9 to 35.6 by 21.3 to 26.0) and averaging 1.24 by 0.95 (31.6 by 24.1 mm) . . . creamy white in color, spotted and blotched with light golden brown.

Egg laying begins in early April in southern California (perhaps even earlier in Baja California), in late April in central California, and in late May or early June in northern California and the Great Basin. After the clutch is completed, incubation takes 22-23 days; so hatching usually occurs about 5 weeks after laying begins. Table 3 summarizes peak periods of laying and hatching as observed in various portions of the California Quail range. As Erwin has shown (Appendix C), the timing may differ substantially from year to year in the same locality. The dates presented in the table are rough averages of the ''normal'' breeding cycle.

Incubation is normally done entirely by the female, but if she has been killed the male may assume the incubation duties (Glading, 1938b; Williams, 1969). In any event, the male is constantly attentive to the female and her nest, and plays an important role in the rearing of chicks.

California Quail are so successful in hiding their nests and so secretive in approaching and leaving them that relatively few have been found by

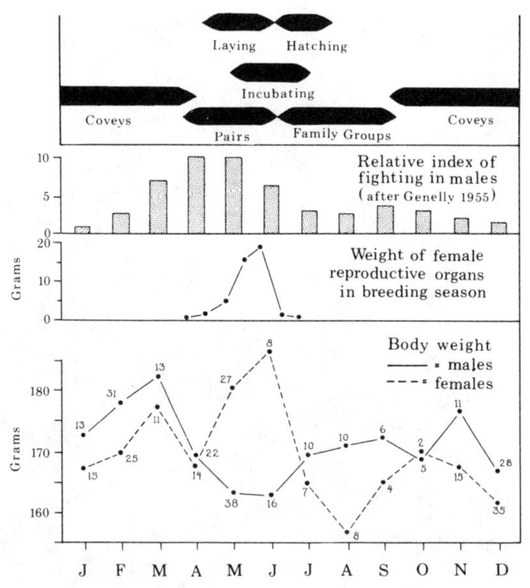

Figure 31. Annual changes in social structure in California Quail with associated changes in average body weight and aggressive behavior. (After Lewin, 1963.)

TABLE 3.
Peak periods of egg laying and hatching in various parts of the California Quail range

Authority	*Area*	*Peak periods of:* *Egg laying*	*Hatching*
Grinnell et al. (1918:529)	California	April–June	May–July
Sumner (1935:217)	Santa Cruz Co.	April–June	May–July
Genelly (1955)	Berkeley, Alameda Co.	April–June	May–July
Lewin (1963:267)	Berkeley, Alameda Co.	Late April–late May	Late May–July
Erwin (Appendix C)	Shandon, San Luis Obispo Co.	April–late May	Late May–July
Glading (1938a:321)	San Joaquin Experimental Range	April–June	May–July
Anthony (1970a)	Southeastern Washington	May–June	June–July
Nielson (1952)	Uintah Co., Utah	May–June	July
Savage (1974)	Modoc County	June–July	Mid-July–August

Figure 32. Nest of the California Quail in mixed annual grasses near the Lick Hills. G. Pickwell photo.

quail investigators. By far the most successful study of California Quail nesting was conducted on the San Joaquin Experimental Range, Madera County, by Glading (1938a). He found 93 nests during the 1937 season, most of them in May and June. Over half of these (50) were situated in dry grass or weeds, the balance in miscellaneous sheltered sites such as rock piles, gullies, at the base of shrubs, etc. Eggs were deposited at the rate of three every four days. Genelly (1955) reports laying hens to average 5 eggs per week. As with other gallinaceous birds, many eggs were dropped in the field, apart from nests. The average complete clutch consisted of about 12 eggs. Of the 93 nests, 40 reached the stage of incubation, and 17 successfully hatched after the normal incubation period of 22-23 days. Figure 34 depicts the fate of the 93 nests. Some pairs that lost an initial nest to a predator or other cause were found to try again. Second clutches were smaller, averaging 8.5 eggs.

Figure 33. A female California Quail incubating her clutch under the shelter of a leaning juniper. M. Erwin photo.

It will be noted in Figure 34 that a substantial proportion of the nests were robbed by predators, 30 of them by the Beechey Ground Squirrel *(Citellus beecheyi)*. On the San Joaquin Experimental Range, at the time of Glading's study, ground squirrels were abundant (10 per acre), which is not typical of most California rangelands, where squirrels are closely controlled. Under present circumstances in most quail ranges of the state, squirrels and other nest predators are not the serious liability on quail nesting that Glading reports. Where grazing is not excessive and adequate nesting cover is available, quail are able to rear enough young to maintain populations commensurate with the carrying capacity of the range. Studies of many North American game birds that nest on the ground have recorded nest losses to predators approximating 50 percent. Presumably these birds, quail included, have a high enough breeding potential to absorb that

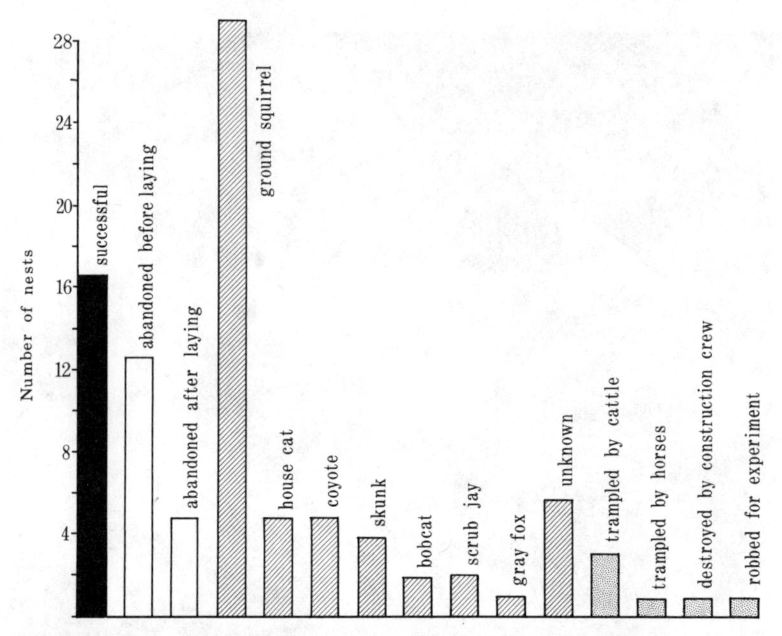

Figure 34. Fate of 96 California Quail nests observed by Glading (1938:333) on the San Joaquin Experimental Range in 1937. Seventeen nests hatched successfully, the rest failed for reasons indicated.

level of loss. Control of nest predators, therefore, is not often a necessary component of quail management.

DOUBLE-BROODING

A pair of California Quail normally establishes a single nest and produces one clutch of eggs. If these are brought successfully to hatching, the two parents devote their energies to rearing the brood. If, on the other hand, the nest is destroyed before hatching, the pair may re-nest and make a second or even a third attempt to bring off a brood. In this event, young chicks may be produced late in the season, when the early hatched birds of other pairs are well grown. The fact that some young are produced in late summer does not of itself constitute evidence that any one pair has produced a second brood.

On the other hand, in some years when the quail show unusually strong sexual drives, there may indeed be cases of double-brooding. McMillan (1964) observed phenomenally high production of young in the Shandon area in 1952, and concluded that some first broods were being reared by the male parent while the female sought a new mate and produced a second brood. The evidence, however, was circumstantial. Subsequently, Francis (1965a) fully documented that double-brooding does indeed occur. In 1963, in a large outdoor pen near Berkeley, eight wild-trapped quail were per-

Figure 35. California Quail nest in the process of hatching. J. B. Cowan photo.

mitted to breed in a normal manner. Two pairs brought off broods of 16 and 17 chicks respectively, but after about two weeks the hens deserted their families, found new mates, and nested again. In each case the male reared the first brood. The females with their new mates produced 9 and 10 chicks respectively in their second broods. All of the adult birds were individually marked, so there could be no question about identity. Anthony (1970a) cites additional evidence of possible double-brooding among California Quail in southeastern Washington.

The phenomenon of double-brooding seems to occur once or twice per decade, in years that for one reason or another are highly favorable for reproduction. Possible causative factors are discussed in Chapter 9.

In years of exceptionally high production, the unmated cocks assume an important function as foster parents. Although lone cocks seem at all times to have a solicitous interest in caring for chicks, they are usually repulsed by jealous parents. When, however, large numbers of chicks are afoot, a certain number are separated from their parents, and these are gathered up and successfully cared for by the bachelor cocks. Whole broods may be taken over by these benevolent "uncles." Erwin (Appendix C) observed this process at Shandon in 1973. He saw two cocks caring for a heterogeneous assortment of 93 young of different ages, hence obviously of different broods.

NON-BREEDING IN UNFAVORABLE YEARS

At the other extreme, there are occasional years in the arid ranges when quail do not breed at all. A. Leopold (1933:28) notes that "during periods of drought, Gambel quail coveys fail to pair off and nest. Apparently in such instances the disposition to breed is inactive for lack of some stimulus associated with normal weather, food, and cover, but the abnormal condition does not visibly affect the health of the adult birds."

In 1950, Macgregor and Inlay (1951) observed such a case of non-breeding in Gambel Quail in the Colorado and Mojave desert region of southeastern California. At the normal season of nesting, the birds did not pair off but remained in winter coveys. Several birds were collected to examine the development of the gonads. Males showed only a slight enlargement of the testes, and there was no visible sign of development in the ovaries or oviducts of females. The year 1950 was unusually dry, and the birds were found to be living exclusively on seeds, there being no green food available. Macgregor and Inlay surmised that lack of greens in the diet deprived the birds of some qualitative dietary element needed for reproduction.

The same phenomenon of non-breeding has been noted in California Quail during exceptionally dry years. Grinnell et al. (1918:528) cite a number of reports of non-breeding. Leopold (1959) notes that non-breeding occurs in the desert regions of northern Mexico in Gambel, Scaled, and California Quail. At the tip of Baja California, rainfall was minimal for the three-year period 1970-72, and the California Quail coveys diminished to a few birds, all obviously survivors of the last period of breeding, which occurred some time in the late 1960's. Heavy and sustained rainfall during the winter of 1972-73 led to successful breeding and partial recovery of numbers.

It is clear, therefore, that in arid and semi-arid regions the reproductive propensity of quail operates on a variable scale, from non-breeding on the one hand to double-brooding on the other. In all cases, rainfall and its concomitant effect on vegetation seems to be the external regulatory factor. The actual physiologic mechanism by which rainfall/vegetation stimulates or inhibits quail reproduction will be considered later. At this point it is sufficient to note that variable reproduction is a highly adaptive characteristic in birds inhabiting arid ranges. In dry years, there probably would be insufficient food to rear young or to winter them, so teleologic reasoning would dictate that the adults save their strength and wait for a year when reproduction would stand a chance of success. When a good year occurs, the birds capitalize on it and restock all the available coverts.

QUAIL REPRODUCTION IN MESIC AND COOL REGIONS

The positive correlation between rainfall and reproductive success, described above for the arid ranges, is reversed in the humid forest ranges and the Great Basin ranges (see Fig. 14). There, warm dry springs lead to maximum reproduction, and cool wet springs severely depress rearing success. Presumably the birds breed every year, but the effectiveness in rearing young is compromised by precipitation during the hatching season.

The best evidence of this relationship is supplied by Savage (1974), who studied a California Quail population in Modoc County, northeastern California. For seven consecutive years (1966–72 inclusive), Savage trapped and banded quail on ranchlands near Alturas, recording the sex and age of each bird handled. He found an *inverse* correlation between the proportion of young birds in the autumn population and the amount of rainfall recorded the previous June. This situation will be further discussed in Chapter 9, but it is mentioned here to signify that the factors affecting nesting success may be very different in diverse parts of the quail range. Speaking of the Modoc uplands, Savage states (p. 4): "The major factor affecting quail reproduction appears to be the amount of precipitation during June that may destroy nests, delay nesting and increase mortality of chicks. Thunder showers are common and often deposit over .50 inches in a 24-hour period."

Sumner (1935), in his study of a quail population in the moist coastal hills near Santa Cruz, noted that quail chicks are highly susceptible to chilling when wet. Raitt and Genelly (1964) confirmed Sumner's observation that wet grass may severely reduce the probability of survival of young quail chicks. As regards their observations of quail in the Berkeley hills, they state (p. 130):

> Although the season when the quail on the study area are hatching and undergoing early stages of development is normally fairly dry, it is occasionally beset with measurable rainfall and heavy, persistent fog, which may last for several days. Under such conditions, newly hatched quail may become soaked and remain so until humidity drops and temperature rises. . . . [T]he effects of persistent fog [and wet grass] . . . are probably beyond the ability of parental brooding to overcome. Additionally, continuous brooding precludes feeding, and the chicks may starve.

The authors then present data showing a general negative correlation between young reared and the number of foggy or rainy days in June and July.

In summary, it is impossible to make a categorical statement about how spring weather affects quail reproduction. In the arid zones, winter and spring rainfall is favorable; in humid coastal areas and cool interior uplands, spring rainfall is highly unfavorable.

7

GROWTH AND DEVELOPMENT OF THE YOUNG

BROODING

For some days after hatching, quail chicks are highly susceptible to chilling and in a lesser degree to overheating. They are unable to regulate their own body temperature without the attention of the hen. The normal procedure is for the hen to brood the young at night under her fluffed-out body feathers. In the morning, when the sun has warmed the ground and dried the dew from the grass, the hen and chicks sally forth to feed. In arid ranges of interior California, this condition is reached shortly after daylight. Along the Pacific Coast, where night fogs are common in summer and where ground vegetation tends to be dense, it is often late morning before the grass is dry enough for the chicks to move about. It may be that one of the important benefits of grazing in coastal vegetation is the exposure of some open ground which permits quail chicks to forage.

After a period of morning feeding, the hen leads the chicks into the shade during the heat of the day. There is a second period of feeding in the afternoon, before the hen gathers her young under her for the night.

The role of the male is to guard the family from danger. He is constantly alert to sound a warning if a predator is observed, sending the hen and the chicks scurrying for concealment, where they remain motionless and ''frozen'' until the cock signals that all is well. On occasion, the cock will attack and drive away intruding animals that threaten the brood. Sumner

Figure 36. A pair of California Quail with newly hatched brood. M. Erwin photo.

(1935:220) recounts attacks by cock quail on meadow mice, a thrasher,a California jay, a roadrunner, a king snake, a spotted owl, and even a dog. Occasionally the female will participate in diversionary attacks against a predator, as exemplified by an attack on a weasel by a hen quail collaborating with an English sparrow. On rare occasions, the female will feign injury to draw a predator away from her brood, but this ploy, commonly used by other birds, is not often observed in quail.

In the event that the hen is killed during the period of egg incubation or brooding of the newly-hatched chicks, the mothering role is assumed by the cock (Grinnell et al., 1918; Glading, 1938b). In those rare years when some quail hens hatch second broods, the male is left to attend the first offspring, but this usually occurs when the chicks are two weeks old, after the critical brooding period is over. Unmated cocks gather up the strays and thereby assist with the rearing.

NUTRITION OF YOUNG CHICKS

A quail chick is born with a substantial residue of the original egg yolk enclosed within the abdomen. This built-in food source serves a vital function in tiding over the bird during the critical period while it learns to gather its own food. It also makes possible the survival of chicks during a period of rain or fog when a day or even two days may pass without opportunity to forage. Lewin (1963) dissected chicks of various ages and depicted the

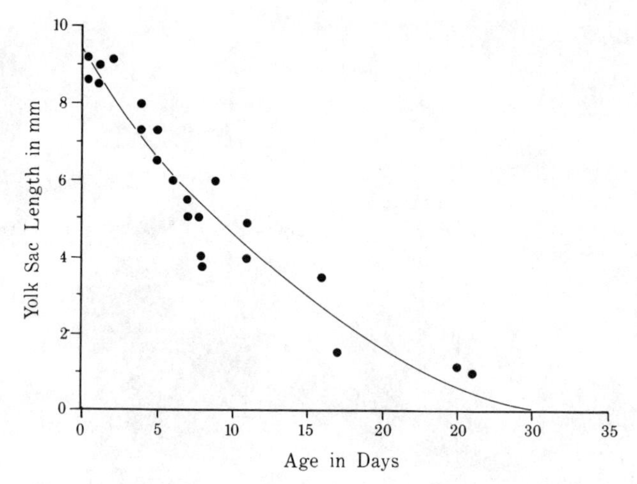

Figure 37. Resorption of the yolk in quail chicks. (After Lewin, 1963.)

rate of resorption of the yolk sac, as shown in Figure 37. By the end of two weeks, the nutriment is largely exhausted, but by then the chicks should be feeding regularly.

Most young gallinaceous birds start foraging largely on insects, with an increasing intake of seeds as they get older. The food habits of the Bobwhite, for example, have been studied intensively. C. O. Handley observes (in Stoddard, 1931:159):

> Young bobwhites on leaving their nest begin feeding at once, though they eat only a small quantity of food during the first 24 hours after hatching, and this consists mainly of fragments of small insects. Grit is taken along with the first food. The chicks either learn by experience or are taught by their parents what they shall eat. A low call uttered by the parent on securing a bit of food brings the whole brood scurrying at times. The first to reach the morsel usually carries it to a place where it can be picked into bits, but at times several chicks grasp the piece of food and tear it apart in "tug of war" fashion.

During the first two weeks, the food of young Bobwhites was found to consist of 83.7 percent animal matter, mostly beetles, grasshoppers, spiders, caterpillars, and bugs. Seeds and fleshy fruits began to assume prominence in the diet of chicks in the third week, after which animal foods dropped rapidly in importance. By late summer the young are living almost exclusively on plant foods, like their parents.

In the case of California Quail, however, available data suggest that chicks are much less dependent on insects than some of their relatives, and that almost from the inception of feeding they learn to recognize and to pick up weed seeds from the ground surface. At the San Joaquin Experimental Range in Madera County, a thorough study was made in 1973 and 1974 of chick food habits by Newman and others (unpubl. data), which showed that the crops of 47 chicks 1–3 weeks of age contained only 10.9 percent insects. In a sample of 66 chicks from 4 to 6 weeks of age, the insect component dropped to 9.3 percent; thereafter it fell rapidly to only a trace. Green leafage constituted only a fraction of one percent of the chick diet, the bulk of which was made up of weed seeds.

Likewise, Anthony (1970's) recorded the food of 29 juvenile California Quail collected in southeastern Washington during the summer of 1967. The contents contained 9 percent insects and 74 percent seeds—a diet not materially different from that of adults taken during the same period. Green leafage made up only 0.7 percent of the summer diet of chicks.

Although insects constitute a small proportion of the food of California Quail chicks, the protein in this dietary component may be quite important for the health and development of the young birds. It is possible that the aerial application of insecticides on croplands and some rangelands may prejudice the survival of young quail by reducing the supply of insects during the first critical weeks of their existence.

THE INTESTINAL FAUNA

Digestion of coarse foods by adult quail, particularly of bulky greens, is made possible by the existence of an intestinal protozoan fauna. The microorganisms live in great abundance in the digestive tract, with highest densities occurring in the caeca—two blind appendages of the gut extending from the point of junction of the large and small intestines. Leopold (1953) reported on the intestinal morphology of 19 species of gallinaceous birds and showed that species that live seasonally on coarse plant foods (such as greens, buds, catkins, etc.) have appreciably larger caeca than those that live mostly on seeds and other food concentrates. The caeca were assumed to be the site of microbial digestion of cellulose, an assumption supported by some experimental work by two Finnish scientists, Soumalainen and Arhimo (1945). It has subsequently been shown that the protozoa also are primary producers of several essential vitamins, including biotin, riboflavin, niacin, and folic acid.

As regards California Quail, Leopold (1953) demonstrated that the caeca of birds taken from a coastal population feeding heavily on greens were significantly larger than those found in a sample of interior birds whose autumn diet was mostly seeds. At the time, this difference was assumed to be genetic, but Lewin (1963) subsequently demonstrated that caecal dimension in quail varies with diet. When birds shift from seeds to greens, as normally happens in winter, the caeca enlarge to hold the bulk food during the time needed for digestion by the protozoa. Lewin identified the principal organisms in the quail caeca as belonging to the genera *Trichomonas, Eutrichomastix,* and *Eimeria*. Without these digestive symbionts, quail could not survive on green roughage.

Quail chicks have to obtain a "seeding" of protozoa from adult birds. This probably happens when the chicks peck at and ingest fragments of droppings from their parents. By the time wild chicks have passed the period of insect and seed dependence and have started to live partially on green foods, they presumably have received the requisite innoculation of protozoa.

RATE OF GROWTH OF YOUNG QUAIL

Several investigators have plotted weight curves for young quail of known age. Lewin (1963:269) states: "The published growth curves of young California Quail are not in close agreement. The 15 birds raised in an outdoor pen by Sumner (1935:225) show a slower rate of growth than do the pen-reared chicks raised by Genelly (1955:281). Genelly also shows that wild birds have a faster growth rate than either of the above pen-reared groups of young and postulates that a freedom of food choice might account for accelerated growth." Lewin proceeded to raise some young quail of his own, part of which received innoculation of intestinal protozoa, and part of which were kept sterile. The plot of his results is presented here in Figure 38. It will be noted that the birds with the intestinal fauna were re-

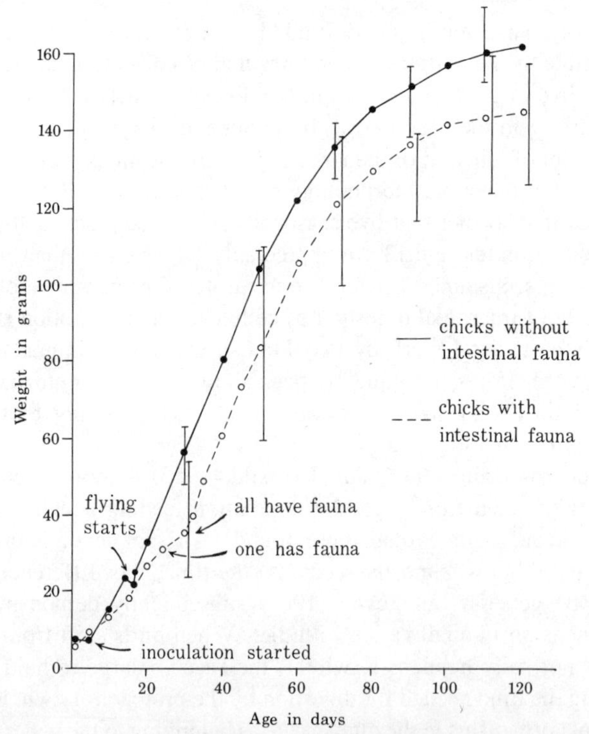

Figure 38. Growth curves of California Quail chicks, with and without normal intestinal fauna. (After Lewin, 1963.)

tarded in growth and at four months of age were 10 grams lighter in body weight than the uncontaminated birds. He comments as follows (Lewin, 1963:271): "The association of caecal protozoans with decrease in growth rate seems to indicate that these micro-organisms have a temporary depressing effect on the rate of development from which the birds never fully recover. . . . Presumably the presence of an intestinal fauna is advantageous to adult gallinaceous birds in helping them to digest coarse foods high in cellulose. However, these organisms quite clearly are something of a burden in the intestinal tracts of chicks, at least as judged by the effect on growth rate."

By the age of four months, young quail approach the size and weight of adults.

BURSA OF FABRICIUS

The bursa of Fabricius is a small lymphatic organ that develops in many kinds of birds during the period of rapid growth and then as abruptly involutes and disappears as the bird matures. The bursa is situated on the dorsal surface of the cloaca. A lumen or central canal opens into the cloaca. The apparent function of this organ is to produce antibodies that protect birds from infections (Farmer and Breitenbach, 1966).

In birds that undergo a complete post-juvenal molt, so that young birds in their first winter cannot be differentiated from adults on the basis of any plumage characteristics, the bursa serves as a useful aging criterion. It is widely used to separate young from adults in ring-necked pheasants, various kinds of grouse, and waterfowl, among others. By probing the depth of the bursal opening, situated just inside the vent, young (with a deep bursal canal) can be differentiated from adults (that have only a shallow pit remaining). This characteristic is not generally used in aging quail because it is so much easier to determine age by a glance at the greater upper primary coverts of the wing. Nevertheless, Lewin (1963:272) studied the growth and resorption of the bursa in young California Quail. Figure 39 shows the bursa of a young bird and the resorbed remnant of this structure in an adult. Figure 40 plots bursa size (width in mm) against age in a series of young birds. The gland grows rapidly in size from hatching (2 mm) to about nine weeks of age (8 mm), whereupon it begins to shrink. Birds taken in the autumn and winter shooting season can still be aged by bursa size.

PLUMAGES AND MOLTS

A quail chick is hatched in natal down and goes through two full plumage molts before it achieves the garb that it will wear through its first winter of life and its first breeding season. Thereafter, as an adult, it will replace its plumage once a year in late summer, after breeding. The sequence of plumages in young quail has been best described by Raitt (1961), and the present account is based on his work.

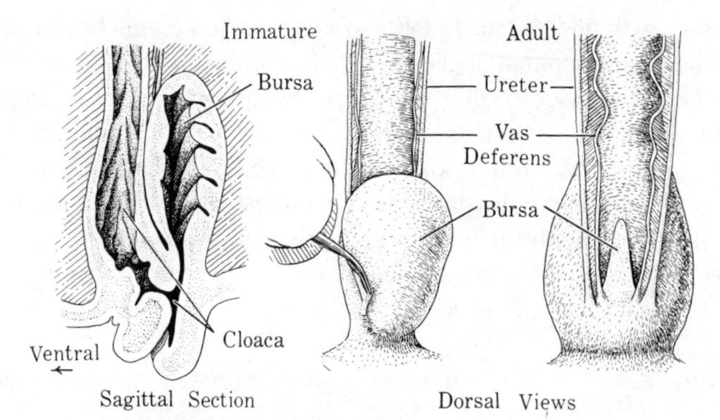

Figure 39. The bursa of Fabricius of young quail, seen in sagittal section (left) and dorsal view (center). In adult birds the bursa atrophies (right). (After Lewin, 1963.)

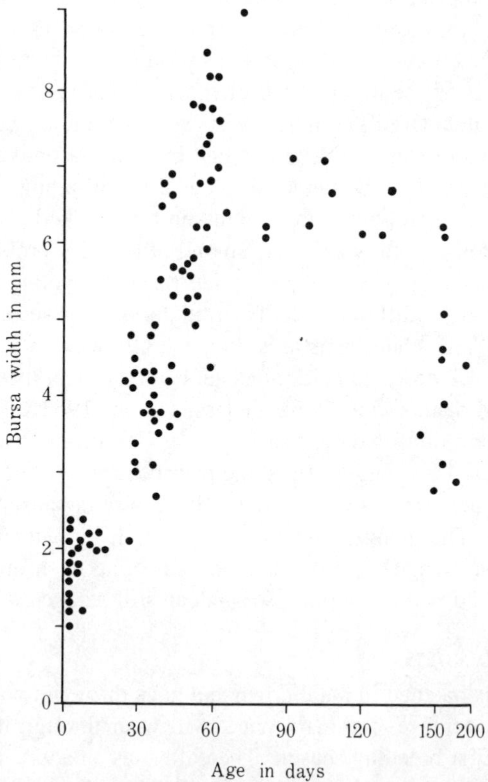

Figure 40. Growth and resorption of the bursa of Fabricius in California Quail chicks. (After Lewin, 1963.)

1. *Natal down.*—The downy plumage of a newly-hatched chick is basically buff-colored with concealing longitudinal streaks of dark brown. Even as the bird emerges from the shell, however, the juvenal plumage is starting to erupt in the form of tiny pin-feathers in the wing, which will become the first flight feathers.
2. *Juvenal plumage.*—The post-natal molt produces a juvenal plumage that is fully developed at about 10 weeks of age. The juvenal feathers are tan with darker brown streaks and bars. There is no sexual differentiation.
3. *First-winter plumage.*—The post-juvenal molt starts in the wing feathers at about five weeks of age, before the juvenal plumage as a whole is fully developed. In this molt the birds assume the appearance of adults, with the sexes clearly differentiated. It is the first-winter plumage that the birds wear until the following summer, when all feathers are replaced in an annual molt. The sequence of feather replacement in the transition from one plumage to the next is best presented in Figures 41 and 43, taken from Raitt's publication.

Figure 41 depicts the growth and replacement of the wing feathers (primaries and secondaries) and the tail feathers (rectrices) plotted against age of the bird in weeks. A quail has 10 primaries, which are numbered from the wrist joint outward. It has 15 secondaries, numbered from the wrist joint inward. The tail consists of six pairs of rectrices, or 12 feathers in all.

Seven of the juvenal primaries are growing at hatching time. Within a week, eight secondaries erupt. These flight feathers grow rapidly, so that the young bird is capable of making short flights by the time it is a little over two weeks of age. The adaptation to early flight contributes impor-

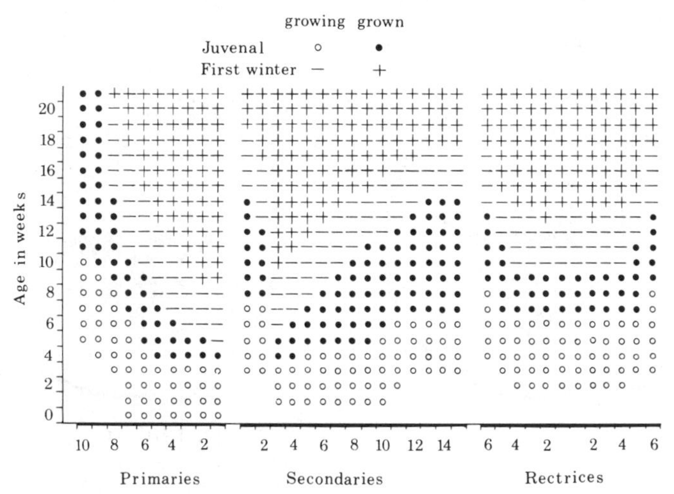

Figure 41. Molt of wing feathers (primaries, secondaries) and tail feathers (rectrices) in California Quail. (After Raitt, 1961.)

tantly to the survival of young quail by minimizing the time that the birds are helplessly exposed to ground predators.

It will be noted that by the age of 21 weeks all the juvenal flight feathers of wing and tail have been replaced by post-juvenal feathers except juvenal primaries 9 and 10, which are retained as part of the first winter plumage. Also retained are the juvenal upper primary wing coverts, which are differently colored from adult coverts and which serve as easily recognizable markers to tell young birds from adults in fall and winter. (Sumner, 1935; Leopold, 1939). Figure 42 shows the coverts as viewed on the spread wings of adult and young birds.

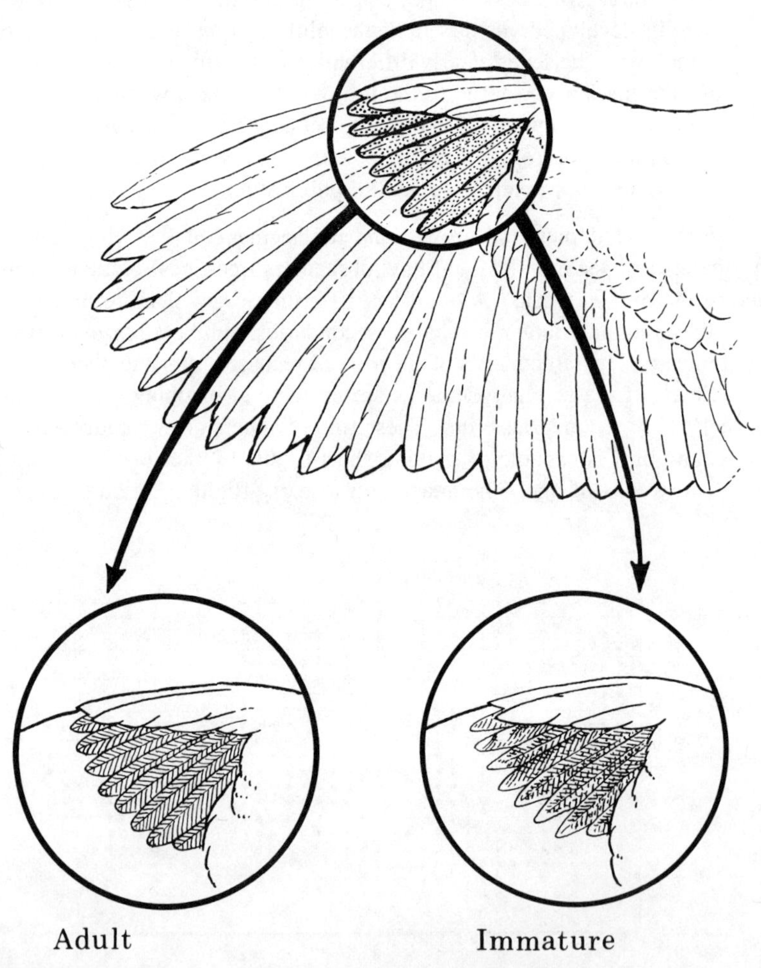

Figure 42. Greater upper primary coverts on the wing of California Quail, used to separate adult from immature birds. Adult coverts are solid gray with rounded tips. Coverts of immature birds-of-the-year are buff-tipped and more pointed in shape.

Figure 43 depicts the stages of body molt in quail, from hatching to the age of 19 weeks. In general, each molt tends to start on the flanks, spreading over the back and across the breast. Parts of the head and the abdomen are the last to molt.

As regards the appearance of distinctive plumages in male and female California Quail, reference to Figures 26 and 28 in the previous chapter will show that the juvenal testes and ovaries respectively begin to approach adult size in August, when the birds are about 8 to 10 weeks old. Sexually dimorphic plumage color in most birds is induced by the presence of sex hormones in the bloodstream. Presumably these hormones are flowing when young quail undergo their post-juvenal molt (7 to 18 weeks of age), thereby accounting for sexually distinctive plumages.

PLUMAGE MOLT IN ADULTS

Adult quail begin to molt their plumage in mid-summer, while the chicks are being reared. The sequence of feather replacement follows essentially that of the post-juvenal molt of young birds except that all feathers are replaced, including the outer primaries and the greater upper primary coverts. Raitt (1961:301) recorded the stage of molt of a number of adult birds handled during the course of his four-year study near Berkeley. His data are assembled in Figure 45. Males begin their annual molt a full month earlier than females, but the rate of feather replacement is slower in males, so both sexes complete the growth of a new plumage at about the same time, in late October. Erwin (Appendix C) has shown that in years when nesting is desultory and is terminated early, molt is initiated earlier than usual, in both sexes. In years of protracted nesting, the molt may be delayed as much as a month.

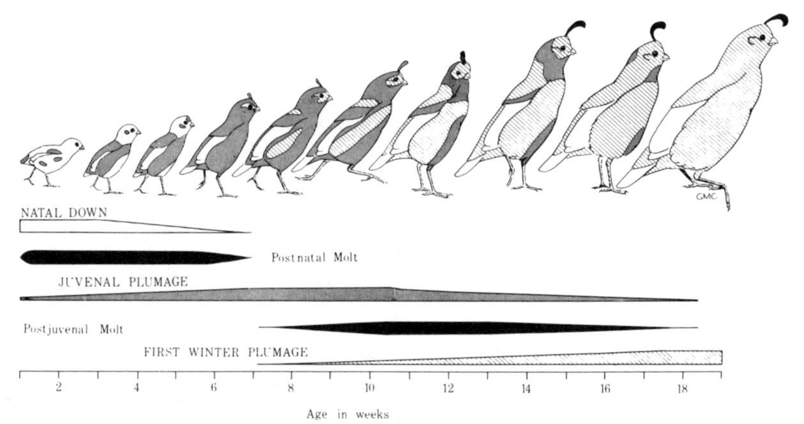

Figure 43. Development of body plumage of California Quail. Stippling indicates areas bearing natal down; vertical hatching, areas covered by juvenal feathers; oblique hatching, areas covered by first winter plumage. (After Raitt, 1961.)

Figure 44. Young California Quail in juvenal plumage, about six weeks of age. Sexual dimorphism is not yet evident. M. Erwin photo.

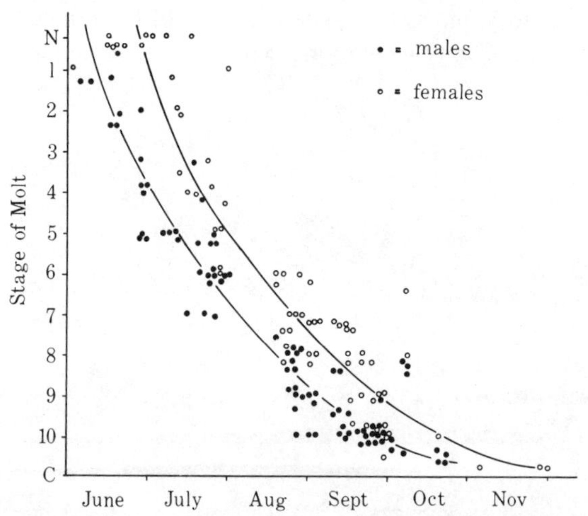

Figure 45. Timing of molt of the 10 primary wing feathers (vertical ordinate) of the California Quail, based on wild birds trapped near Berkeley over a 4-year period. (After Raitt, 1961.) For evidence of molt irregularity, see Fig. 93, Appendix C.

One peculiarity of molt in quail, as in many other birds, is the production of a "brood patch" in females during incubation of the eggs. The feathers of the abdominal region drop out and the skin becomes edematous and highly vascular, which facilitates warming of the eggs by the setting hen. Jones (1968) demonstrated that brood patch formation in female California Quail is induced by the hormone prolactin, one of the ovarian secretions which also induces broody behavior. After the chicks are hatched, the abdominal skin returns to normal and new feathers are produced, obliterating the naked brood patch.

SEXUAL MATURITY IN CALIFORNIA QUAIL

Both sexes of quail are fully mature and capable of breeding at the age of 10 months, when they enter their first breeding season. There is, however, a lag in the sexual development of yearling females, which approach the nesting period with smaller ovaries than adults (see Fig. 28, Chap. 6). The size differential disappears when ovulation begins, but there remains a serious question whether young females are as efficient and successful as adults in producing young. Francis (1970) concluded that they were not. The proportion of the adult females in a population seems to be an important factor affecting productivity in any given year: the higher the ratio of adults to young, the higher the production per female, other factors being favorable. Francis states (p. 258):

> In both the Shandon and Berkeley areas . . . every case of abnormally high reproductive success followed a year in which reproduction was lower than normal. This may have resulted from increased reproduction due to lower population densities; from higher reproductive rates in older females as compared to those in their first breeding season; or from a combination of these effects. Smaller populations, following a year of poor reproduction, have a higher proportion of old birds in the following season's breeding population. Observations of the birds in captivity suggest that the breeding efficiency of older females is more important than density. In the 2 years in which breeding was observed in Berkeley, with essentially no difference in population density, four adult females hatched 83 chicks in six nests, while three yearling females failed to hatch any chicks at all.

This question will be further explored in the next chapter.

There is no evidence that young males are any less virile than adult males, although they may be lower in the social pecking order and less likely to obtain a female for a mate (Genelly, 1955:268).

8

SEX AND AGE RATIOS AND THEIR INTERPRETATION

SOURCES OF COMPOSITION DATA

A population of quail is a dynamic and ever-changing aggregate of individuals. Birds die and others are recruited to take their places. A covey may be likened to the leaves on a live oak tree: foliage is always present, but the leaves seen this year are only in part the leaves that clothed the tree the year before. In the technical wildlife literature, the annual rate of replacement of individuals in a population is referred to as "turnover."

There are various ways in which turnover rate in a quail population can be measured. The most precise is to trap and mark a substantial cohort of the population and to follow the disappearance of marked individuals and their replacement by unmarked young. From such a record, taken over a period of years, one can compute average annual "turnover rate" and likewise "turnover period," or the number of years required for a marked cohort to be completely replaced. The process of trap-mark-release yields a wealth of data, but it is tedious and time-consuming.

Much of the same information on replacement can be obtained more easily from annually sampling the population by shooting, and ascertaining the proportion of cocks/hens and adults/young in the sample. As noted in the last chapter, adult and immature California Quail (under one year of age) can be recognized easily by examination of the greater upper primary coverts of the wing. From the sex and age ratios so derived much can be learned about year-to-year variations in reproductive success and in mortality. These data in turn can be correlated with annual differences in weather and in vegetation to derive inferences about the forces that regulate population dynamics.

Our records on the make-up of California Quail populations derive largely from samples shot during the legal hunting season. Obviously, this means that we know quite a bit about autumn populations and relatively little about composition during the rest of the year. Interpretation of the yearly cycle depends on extrapolating the data from one autumn to the next. A basic question that must be answered is whether shooting records represent a true sample of population composition.

Considering the latter point first, there is considerable evidence that shooting of quail is generally non-selective and hence that bag samples do indeed portray actual composition. Stoddard (1931:92) showed that shooting records of Bobwhites in Georgia gave approximately the same sex ratio as trapping records obtained concurrently. Emlen (1940) concluded that hunters' kill-records accurately sampled sex and age groups in California Quail. Williams (1957) felt that trapping was a valid method of sampling California Quail populations in New Zealand. In a 9-year study of Scaled Quail in New Mexico, Campbell et al. (1973) found that age ratios based on trap samples and on hunting samples were "virtually identical."

The question sometimes arises whether young gallinaceous birds may be more vulnerable to shooting than adults, especially early in the hunting season. If so, there would be distortion of composition data based on hunters' kill. To test this hypothesis I separated the kill records for each of 11 years of hunting at Shandon (McMillan data, Table 8) into four quarters, representing portions of the three-month shooting season. When sex and age ratios for the separate quarters were computed, there was no statistical difference between quarters in composition data for either sex or age. Figure 46 presents the quarterly analyses of age ratios. If distortion occurs it must be very slight.

In the absence of contrary evidence, I conclude that either trapping or shooting will yield representative population samples, and I use both types of data interchangeably, although I have relatively less of the former.

THE ANNUAL CYCLE

Mortality in a quail population goes on year-round, whereas recruitment is limited to the brief period of reproduction in spring and summer. A local population, therefore, undergoes an annual cycle of numbers, reaching it highest point when the flood of young birds joins the population in summer and its lowest point when the winter survivors enter the breeding season in spring. Even in a relatively stable environment, the number of birds is never static but is constantly undergoing change, either up or down. Figure 47 illustrates three consecutive annual cycles in the population of Cali-

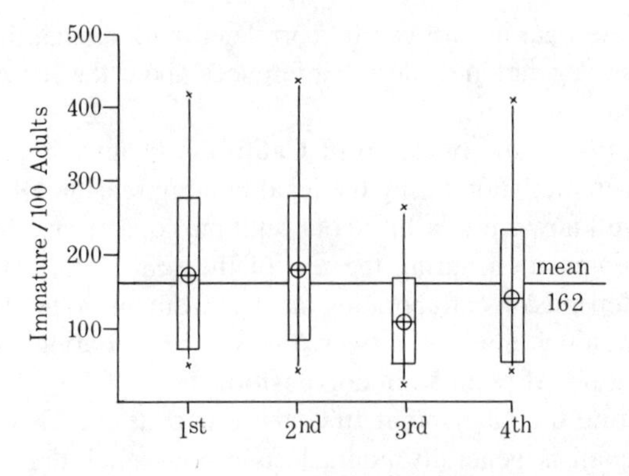

Figure 46. Age ratio of quail taken at various periods (quarters) of the hunting season at Shandon, over the years 1963–1973 (from Table 8).

fornia Quail studied intensively by Emlen (1940) in the vicinity of Davis. As can be seen, reproductive success was high in 1936, moderate in 1937, and low in 1938. The vertical bars on the figure represent populations in November of each year, divided into the adult component (survivors of the past breeding season) and the immature component of young birds that had been recruited. To gather the data for this figure, Emlen trapped and marked most of the individuals in the population for three years, and moreover he counted the total number of birds periodically.

A field study of equal intensity and even longer duration was conducted by Raitt and Genelly (1964) in the hills behind Berkeley. For the six year period 1950 to 1955 inclusive, the quail on a 100-acre study area were captured, marked, and recaptured, and the winter coveys were counted. From data so derived, they were able to construct the life tables presented here in Table 4. The "composite time-specific" table was computed from the age structure of the fall populations. This table says in essence that of 245 young quail (2 months old) alive September 1, 52 would still be alive a year later, 12 the year after that, and so on. Other columns depict the numbers that died during the yearly intervals and the rates of mortality. The "composite dynamic" table expresses the same processes, based on following from year to year the fate of specific marked cohorts of the population.

From the composite mortality rates can be computed an *average* rate of mortality (or average turnover rate), which figures to be 74 percent in the time-specific table and 71 percent in the dynamic table. The difference arises from the fact that the two tables were derived from data from different years.

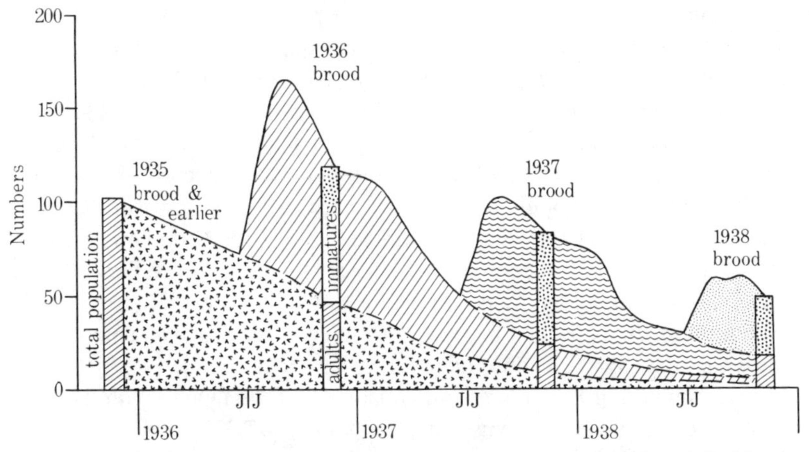

Figure 47. Annual cycle of the quail population on the University of California Davis campus, 1935 to 1938. (After Emlen, 1940.)

TABLE 4.
Time-specific life table for California Quail based on age structure in the fall of 1953, 1954, and 1955 and dynamic life table based on the fate of cohorts of young birds added to the population in 1950, 1951, and 1952 (After Raitt and Genelly, 1964:132.)

	Composite time-specific			*Composite dynamic*		
Age in months	l_x	d_x	q_x	l_x	d_x	q_x
2	245	193	78.8	176	146	83.0
14	52	40	76.9	30	17	56.7
26	12	9	75.0	13	10	76.9
38	3	2	66.7	3	2	66.7
50	1	1	100	1	1	100
62	0	—	—	0	—	—

l_x = number of survivors
d_x = number dying during the 12-month period
q_x = mortality rate during the 12-month period

Figure 47 presenting Emlen's records and Table 4 summarizing the findings of Raitt and Genelly are simply different ways of depicting the processes of mortality and recruitment which determine the annual cycle in populations of California Quail.

SEX RATIO

In wild populations of many species of quail there nearly always is found an excess of males over females. This accounts for the presence of unmated cocks during the breeding season. Table 5 presents a sample of recorded sex ratios of four species of North American quail. In every sample, males outnumber females in ratios varying from 107:100 to 122:100.

It was first noted by Emlen (1940) that when a sample of California Quail is separated into four sex and age groups (adult ♂♂, adult ♀♀, immature ♂♂, and immature ♀♀), the distortion of sexes occurs largely in the adult cohort of the population. The sex ratio of young birds in their first autumn of life (immature) is approximately 100:100. Leopold (1945) observed the same phenomenon in Missouri Bobwhites. Table 6 presents data on sex ratios among adult and immature Bobwhites and California Quail. Among immatures the sex ratio is close to equality, whereas among adult birds there always is a substantial excess of cocks.

In comparing Emlen's records for the California Quail with the Bobwhite data from southern Missouri, Leopold (1945:33) stated: "[I]n both species distortion of the sex ratio is confined to the adult segment of the population . . . [T]here is no differential mortality among young birds in the wild up to the end of December. More hens than cocks must be lost at some point later in the life cycle, and I am inclined to favor Emlen's in-

TABLE 5.
Aggregate records of sex ratios in four species of North American quails

	Area	*Source*	*Sample*	*Sex ratio cocks: 100 hens*
Bobwhite	Georgia/ Florida	Stoddard (1931:90)	19,423	114:100
	Southeast	Stoddard (1931:90)	10,707	120:100
	Missouri	Leopold (1945)	43,819	112:100
	Texas	Lehmann (1946)	1,060	115:100
California Quail	California	Emlen (1940)	15,728	112:100
	Berkeley	Raitt and Genelly (1964)	847	107:100
	Shandon	McMillan data–1949 to 1973	15,166	113:100
	Marysville	Conway data–1971 to 1975	4,869	122:100
	California	This study— 1971 to 1973	7,792	118:100
Gambel Quail	New Mexico	Campbell and Lee (1956)	740	111:100
Scaled Quail	New Mexico	Campbell and Lee (1956)	384	108:100

terpretation that the principal differential loss occurs during the breeding season in both California quail and bobwhites.'' The inference was made that the physiologic stress of reproduction, combined perhaps with predation losses of incubating and brooding birds, led to the higher mortality of breeding females.

However, Williams (1957), working with a wild population of California Quail in New Zealand, reached a contrary conclusion. He trapped a substantial sample of young birds and sorted them as to age classes by the stage of wing molt. Birds older than six weeks could be sexed by the color of body plumage emerging in the post-juvenal molt. When he compiled his data, the proportion of males to females was found to increase steadily

TABLE 6.
Sex ratios of adult Bobwhites and California Quail that have gone through at least one breeding season, compared to ratios of immatures approaching their first breeding season

	Source	*Numbers* Cocks	Hens	*Sex ratio* *Cocks:100 hens*
ADULT BIRDS				
Bobwhite (Missouri)	Leopold (1945)	234	145	161:100
California Quail	Emlen (1940)	298	213	140:100
California Quail	Raitt & Genelly (1964)	151	104	145:100
California Quail	McMillan, data from Shandon, 1949 to 1973	3551	2733	130:100
California Quail	Conway, data from Marysville, 1971 to 1975	1123	674	167:100
California Quail	This study (State), 1971 to 1973	1837	1339	137:100
IMMATURE BIRDS				
Bobwhite (Missouri)	Leopold (1945)	634	620	102:100
California Quail	Emlen (1940)	406	408	100:100
California Quail	Raitt & Genelly (1964)	139	148	94:100
California Quail	McMillan, data from Shandon, 1949 to 1973	4493	4389	102:100
California Quail	Conway, data from Marysville, 1971 to 1975	1552	1520	102:100
California Quail	This study (State), 1971 to 1973	2376	2240	106:100

as the birds matured, implying that differential mortality starts at an early age and continues to maturity. His data are summarized in Table 7.

I have no explanation for the apparent contradiction between Williams' data and our own. His distorted sex ratio in young birds over 18 weeks of age (57.2 percent males, or 133 ♂ ♂ :100 ♀ ♀) is completely at variance with our finding of substantial equality in the sexes at that age, based on samples of thousands of birds, taken over a number of years.

There is one test that can be applied to our own data that may shed light on the origin of distorted sex ratios. I have noticed for some years that the degree of distortion between the sexes in adults was greatest in years of high production of young and lowest in years when few or no young were recruited into the population. When the cumulative data from Shandon were analyzed to test this hypothesis, Figure 48 was derived. With the exception of one year, there seems to have been a general trend toward an increasing excess of adult cocks over adult hens as the total productivity for the year increased (as measured by the ratio of immatures to adults in the autumn/winter population). I interpret this to mean a proportionately higher mortality of adult hens in years of intensive and effective breeding effort. The one year that does not conform to the above curve (1952) was a year of prodigious reproduction when McMillan noted that many cocks were taking over the duties of rearing the early-hatched broods of chicks. Perhaps that year the cocks were exposed to the responsibilities and dangers of "motherhood" at a level ordinarily experienced only by females. Be that as it may, and ignoring the aberrant year, the relationship between reproductive effort and hen mortality seems real enough. These data are offered in substantiation of the idea that differential mortality of the sexes occurs largely in the breeding season and probably derives from the greater strains and risks imposed on nesting females.

To sum up, in most samples there appears to be an almost even sex ratio among California Quail in their first winter of life and a distorted sex ratio in adults that have gone through one breeding season. The more effec-

TABLE 7.
Changing sex ratio of young California Quail trapped in the wild in central Otago, New Zealand (after Williams, 1957)

Age group	*Sample*	*Males*	*Females*	*Percent males*
6–9 weeks	50	14	36	28.0*
10–13 weeks	172	62	110	36.1*
14–17 weeks	190	103	87	54.0
18+ weeks	42	24	18	57.2
Adults	299	192	107	64.0

*No explanation is forthcoming for the apparent high preponderance of females in the samples of young birds.

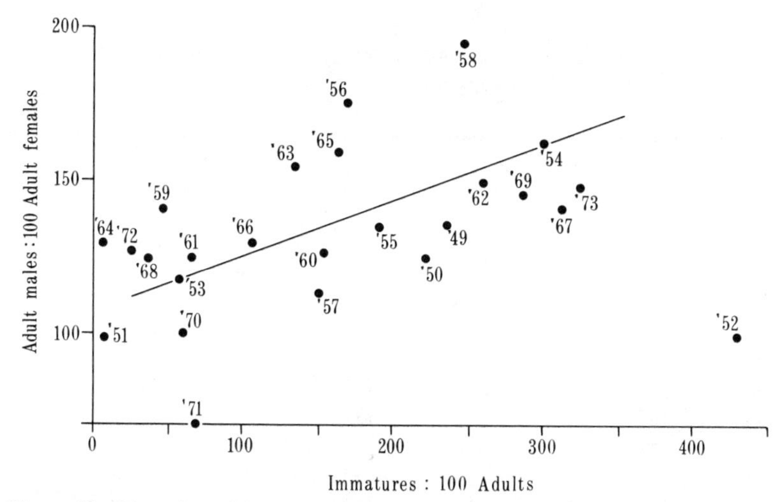

Figure 48. Distortion of the sex ratio in adult quail in relation to productivity during the preceding breeding season. The higher the production of young, the lower the survival of adult females (data from Table 8).

tive the breeding season, the greater the distortion in favor of cocks.There are some data from New Zealand that suggest that distortion in favor of cocks may at times occur even in immature birds. In any event, the resultant sex ratio in winter populations always shows an excess of males.

AGE RATIO

By far the most extensive record of population composition in California Quail is that compiled by McMillan in the vicinity of Shandon, San Luis Obispo County. Some of these data were published in McMillan's paper of 1964. Table 8 brings the record up to 1973. In the 25-year period, a total of 15,166 quail were individually recorded as to sex and age.

Age ratios for each year are expressed in Table 8 as the number of immatures per 100 adults. Reproductive success at Shandon is enormously variable, oscillating from a low of 4 immature: 100 adults to a high of 430 immature: 100 adults. As noted previously, Shandon is situated in the arid foothills of the Inner Coast Ranges, and year-to-year variation in rainfall seems to underlie the extreme shifts in breeding success. The mean age ratio for the 25-year period is 141 immature:100 adults, which signifies an average turnover rate of 59 percent per year.

Records of age ratios derived from other portions of the California Quail range, though of shorter duration than the Shandon data, permit regional comparison of average rates of turnover. There appears to be a higher turnover in quail populations in northern areas than in the more southerly and more arid portions of the range. On Vancouver Island, Barclay and Bergerud (1975) recorded an average turnover rate of 77 percent. In Modoc

TABLE 8.
Sex and age distribution of 15,166 California Quail shot in 25 hunting seasons at Shandon, California (Data from Ian McMillan)

Year	*Adult males*	*Adult females*	*Immature males*	*Immature females*	*Totals*	*Immature per 100 adults*
1949	80	59	164	158	461	232
1950	65	52	127	130	374	220
1951	178	180	16	13	387	8
1952	87	86	365	377	915	430
1953	176	149	94	90	509	57
1954	140	86	333	339	898	300
1955	185	136	302	309	932	190
1956	151	86	203	198	638	170
1957	149	131	214	207	701	150
1958	157	80	311	272	820	246
1959	205	145	86	71	507	45
1960	116	91	157	161	525	154
1961	192	152	119	105	568	65
1962	195	131	431	416	1173	260
1963	250	161	292	264	967	135
1964	167	129	6	5	307	4
1965	178	111	239	231	759	163
1966	138	105	147	110	500	106
1967	85	60	218	234	597	312
1968	166	133	57	51	407	36
1969	75	51	192	168	486	286
1970	120	120	66	77	383	60
1971	120	171	93	134	518	78
1972	86	67	14	25	192	25
1973	90	61	247	244	642	325
Totals:	3551	2733	4493	4389	15,166	141 (average)

County, the records compiled by Savage (1974) also indicate a 77 percent turnover. The Berkeley data reported by Raitt and Genelly (1964) show a 71 percent to 74 percent turnover. Emlen's (1940) study of quail near Davis recorded a 67 percent rate of turnover or, as he calls it, replacement. Ray Conway turned over to me kill records for the years 1971 to 1975 derived from his properties in the foothills east of Marysville. Average turnover was 63 percent. By comparison, the 59 percent turnover in quail at Shandon is low. Table 9 summarizes these regional records. Although the records are scanty and in many ways inadequate, they do suggest that in the more northerly and more humid portions of the range the rates of turnover are higher.

TABLE 9.
Range of age ratios and rates of turnover in quail populations sampled for 4 or more years in various portions of the California Quail range

	Area	*Years sampled*	*Number*	*Range of age ratios (immatures: 100 adults)*	*Average turnover rate*
SW Canada					
Barclay and Bergerud (1975)	Vancouver Is.	1969–73	1112	244–429	77%
Great Basin					
Savage (1974)	Modoc Co.	1966–72	1889	106–620	77%
Humid Coastal California					
Raitt and Genelly (1964)	Alameda Co.	1950–57	847	56–189	71%–74%
Sacramento Valley					
Emlen (1940)	Yolo Co.	1935–38	948	85–235	67%
Sacramento Valley Foothills					
R. Conway (pers. comm.)	Nevada Co.	1971–75	4869	123–214	63%
Arid southern California	San Luis Obispo Co.				
McMillan (Table 8)		1949–73	15,166	4–430	59%

As stated above, turnover is a function of mortality and replacement. But which is cause and which is effect? Does high mortality lead to high replacement, or is it the other way around—high production being followed by high mortality? I favor the latter interpretation. Stated simply, the implications of the data presented in Table 9 seem to be as follows. In the more northern and more humid areas, the quail breed with some sucess every year, although reproduction is by no means always the same. Regular breeding leads to consistently high turnover. In arid zones breeding is more sporadic, and whereas in one year there may be an explosion of young birds, the next two or three years may be characterized by little production and corresponding low mortality. Samples drawn from a series of such years will show a lower *average* rate of reproduction and turnover in the arid zones.

This general relationship is probably true in other species of quail whose ranges span humid and arid zones. Average turnover of Bobwhites in the eastern portions of the United States is in the range of 75 to 82 percent (Rosene, 1969:385). In Missouri it was found to be 73 to 84 percent

(Leopold, 1945). In Kansas the average in one locality was 72 percent (Robinson, 1957:30). In the more arid ranges of south Texas the average is 60 percent (Lehmann, pers. comm.).

YEAR-TO-YEAR DIFFERENCES IN PRODUCTION

As part of the continuing study of California Quail populations sponsored by the California Academy of Sciences, Erwin (Appendix C) organized the collection of quail wings from various parts of the state during three hunting seasons—1971–72, 1972–73, and 1973–74. To illustrate the effects of rainfall on regional production, he presents maps (pp. 262–263) depicting the proportion of young per 100 adults for the unusually dry year 1972–73 and the unusually wet year 1973–74. The maps show graphically that production of young quail is moderate to good in central and northern California each year, irrespective of rainfall. Conversely, in the arid ranges of southern California and Baja California there was little reproduction in 1972 but a virtual explosion of young birds in 1973—a year of good rainfall. The decided year-to-year fluctuations in breeding success at Shandon, shown so clearly in the 25-year record depicted in Table 8, are characteristic of other localities in the arid and semi-arid quail range. Chapter 9 considers the ecologic mechanisms that may bring about reproductive differences in response to the variable of weather.

MAXIMUM AGE OF QUAIL

The life tables of Raitt and Genelly (Table 4) illustrate that mortality in a given age class of quail is highest in the first year of life, and that the probability of survival of an individual bird increases slightly with age. Still, in a viable population with a fairly high rate of turnover, the predictable life span of any one member of the population is fairly short. Raitt and Genelly (1964:133) point out that actuarial data on newly-hatched chicks

Figure 49. A lone cock attending a troop of adopted orphans in the boom year of 1973. M. Erwin photo.

are almost impossible to obtain, but after September 1, when the young birds at Berkeley were approximately 2 months of age, the probable life span of the average individual was computed to be 9.7 months. Their tables likewise show that the "turnover period" of any given year class of birds is about five years—in other words, only one bird in several hundred will live to be five years old. They captured one male, however, that was over 80 months old (6½ years), and state that this record of survival must approximate the potential natural longevity of the species. I collected a banded male near Bitterwater, San Benito County, that had been marked as an adult five years previously. The bird therefore was over 6 years old. This likewise may be viewed as something of a world's record performance for staying alive.

How is it, then, that in desert situations, where enough rainfall to induce breeding may not come for several consecutive years, a breeding stock of quail can still persist? As noted earlier, near La Paz, Baja California, there was no known reproduction of quail during the prolonged drought from 1968 to 1973, yet a few very small bands of quail were found in January, 1973, frequenting mesquite thickets near the city. Apparently, at extremely low density, experienced adult quail may persist for several years in certain highly favorable habitats with virtually no mortality. Death rate would appear to be an inverse function of density, reflecting the same thought that was expressed in discussing turnover rate—namely, that mortality is induced by reproduction. When there is no reproduction for several consecutive years, mortality may drop almost to zero.

9

RAINFALL AS A FACTOR AFFECTING REPRODUCTIVE SUCCESS

RAINFALL AND QUAIL REPRODUCTION IN NORTHERN AND COASTAL RANGES

The point has been made in previous chapters that winter and spring rainfall may have salutary or adverse effects on quail reproduction, depending on the area in question. By and large, rainfall during the nesting season is adverse to reproductive success in the cooler and more mesic portions of the quail range. I already have mentioned the work of Savage (1974), who studied the California Quail population in the vicinity of Alturas, Modoc County. From 1966 to 1972, Savage trapped, marked, and released quail during snowy winter weather when the birds were forced into barnyards to find food. Age ratios were determined from the winter samples of birds handled. Figure 50 shows the relationship between percentage of young in a winter population and the amount of precipitation falling during the previous June. The inverse relationship is abundantly clear—the more June rainfall, the smaller the hatch. Savage (p. 4) offers the logical explanation that precipitation during the nesting period ''apparently destroys nests, delays nesting and increases mortality of chicks.''

In the vicinity of Berkeley near the coast, Raitt and Genelly (1964) reported a similar negative effect of rain or persistent fog on the quail hatch. Heavy rains in spring delayed the onset of breeding and reduced breeding success. Fog and summer drizzles during the hatching period further re-

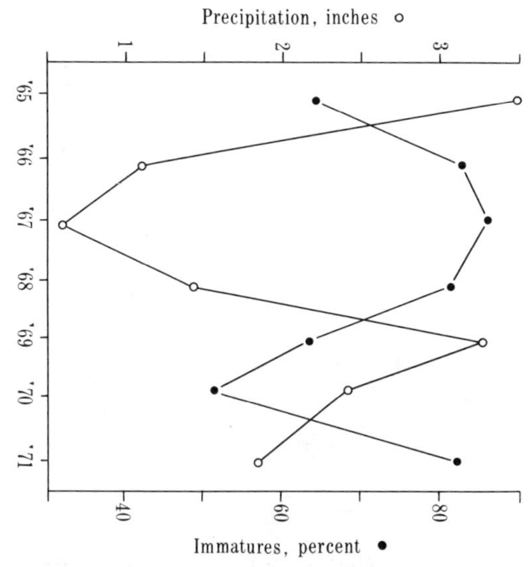

Figure 50. The inverse relationship of California Quail productivity to June rainfall in Modoc County. (After Savage, 1974.)

duced the production of young birds. Mild, dry springs and sunny summers led to the highest productivity.

In the Sacramento Valley foothills east of Marysville, Conway (pers. comm.) finds relatively little difference in production between wet and dry years. In the five year period 1971 to 1975 inclusive, the ratios of immature birds per 100 adults were 207, 123, 140, 214 and 174. The uniformity of these age ratios is surprising, considering that the period included a very dry year (1972) and a very wet year (1973). This area would seem to be the most consistently favorable habitat for quail reproduction of any that has been studied.

RAINFALL AND QUAIL REPRODUCTION IN ARID ZONES

In the arid portions of the quail range, productivity is definitely stimulated by high rainfall, and here the causative mechanism is far less clear. The southern half of California and all of Baja California are in a Mediterranean climatic zone, where rainfall occurs principally in winter. When winter rains are generous, the countryside is lush with spring vegetation. Wildflowers bloom profusely, and many or most species of desert mammals, birds, reptiles, and amphibians reproduce prolifically. It is quite taken for granted that a wet year on the desert is a boom year for plants and animals alike. As a classic example of the direct correlation between winter rainfall and quail reproduction, I reproduce here a figure published by Gallizioli (1965) reflecting his observations of productivity in Gambel Quail near

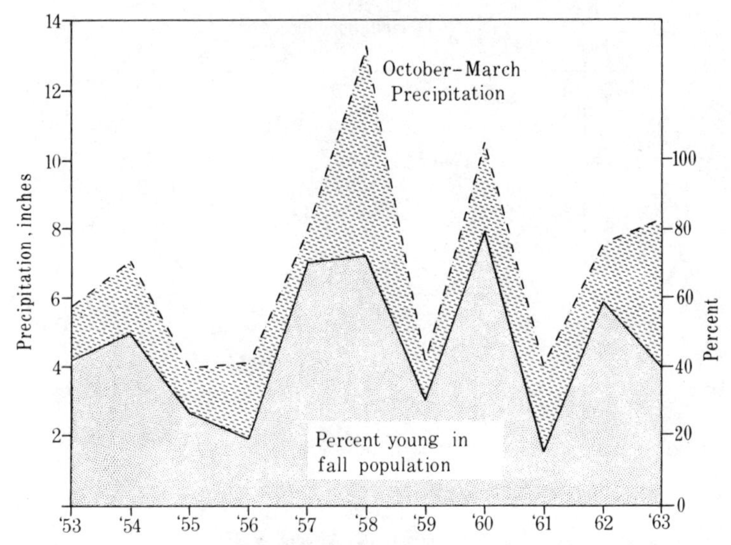

Figure 51. Correlation of winter rainfall with Gambel Quail productivity in Arizona. (After Gallizioli, 1965.)

Oracle Junction, in the desert of southern Arizona (Fig. 51). I presume that if similar data were available for a population of California Quail in the deserts of Baja California they would show the same correlation.

The relationships between rainfall, soil moisture, and plant growth seem logical enough. But the mechanism by which plant growth stimulates reproduction in animals, such as the quail, is by no means obvious.

The data accumulated by McMillan on year-to-year reproductive success of quail in the semi-arid hills near Shandon can be analyzed to seek relationships with weather records. In 1964, McMillan published his records for the period 1950–62 and showed graphically that the years of high quail production corresponded in a general way with years of high total rainfall. McMillan (1964:705) stated: "[T]he data show the irregular but general correlation between seasonal rainfall, quail reproduction, and population trend . . . [F]or all these 5 driest years, with an average of 6.40 inches of rainfall, the average proportion of young in birds bagged was 35 percent. For the other eight seasons, all of which had 8.75 inches of rain or more, with an average of 11.72 inches, the average proportion of young was 70 percent."

But statistically the correlation was by no means a straight-line relationship. There were unexplained irregularities that called for more detailed analysis. Subsequently, Francis (1967) undertook such an analysis. With the aid of a digital computer, he constructed a population model which accurately reflected the year-to-year fluctuations in sex and age composition reported in McMillan's data. The ratio of immature to adult birds was closely correlated ($r = 0.974$) with the estimated changes in population de-

Figure 52. California Quail range east of Ensenada, Baja California. Though badly overgrazed, the area produced a profusion of annual forbs following good rains in 1973 and 1974. A covey of 500 quail was found in this basin on January 12, 1976.

rived from the model. Weather data for the Shandon area were then broken down into 10-day units for the period September 2 to April 29 of each of the 13 years in the sample, and a figure for mean temperature and total precipitation was derived for each 10-day unit. Following a procedure of multiple linear regression analysis, Francis tested a number of weather parameters to determine what combinations of seasonal precipitation and temperature showed positive correlation with the observed ratio of immature to adult birds in a given year. His methods are explained in a paper entitled "Prediction of California Quail Populations from Weather Data" (Francis, 1967).

Simply stated, his findings were as follows. The most important factor relating weather to quail production seemed to be the calculated value of residual soil moisture at the end of April. From cumulative rainfall which was presumed to have entered the soil, a figure for evapotranspiration was subtracted, derived largely from temperature data according to a formula proposed by Thornthwaite and Mather (1957). The resultant computed figure for residual moisture in the soil on April 29 accounted for 83 percent of the variance in the reproductive index. The second most important factor affecting productivity in any given year was the proportion of adult female quail in the population: the higher the ratio of adult to young females, the higher the productivity. This factor accounted for another 13 percent of

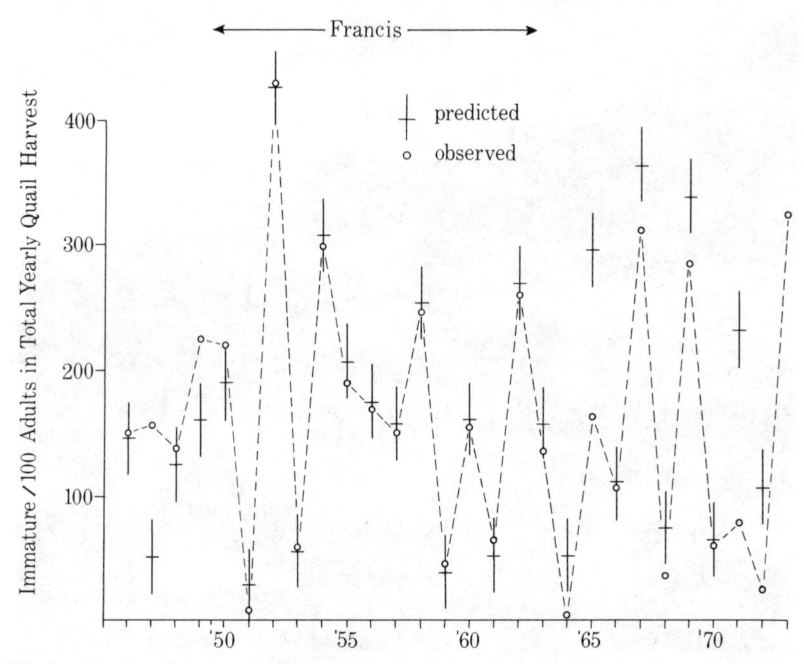

Figure 53. Predicted and observed productivity of California Quail at Shandon, utilizing the predictive formula derived by Francis (1967) from age-ratios and weather data for the years 1950–1962.

the variance. Lastly, and of considerably less significance in the equation, was the total seasonal rainfall for that year (3 percent). These three factors together accounted for 99 percent of the observed variance in reproduction. The central portion of Figure 53 compares the observed ratios of immature to adult quail for the years 1950 to 1962 with the predicted ratios based on Francis' regression equation. The remarkable precision of Francis' formula, based on the 1950–62 weather data and quail data, is evident in the figure.

However, when the formula is used to predict the quail crop in years preceding and following the period from which the formula was derived, the precision breaks down. As shown in Figure 53, predictions for the years 1946–49 and 1963–72 display some marked irregularities. The situation is presented even more graphically in Figure 54, where predicted vs. observed age ratios are plotted in relation to a 45° line, which would represent perfection. Even so, in 18 of the 27 years the predicted age ratios fall within the 95 percent confidence limits of observed ratios, and in several other years they nearly do. In the years 1947 and 1949 the predictions were below the actual ratios, and in 1965 and 1971 the predictions were too high. There may be some significance in the fact that in the early years departures from accuracy tended to underestimate production, whereas in the last 10 years

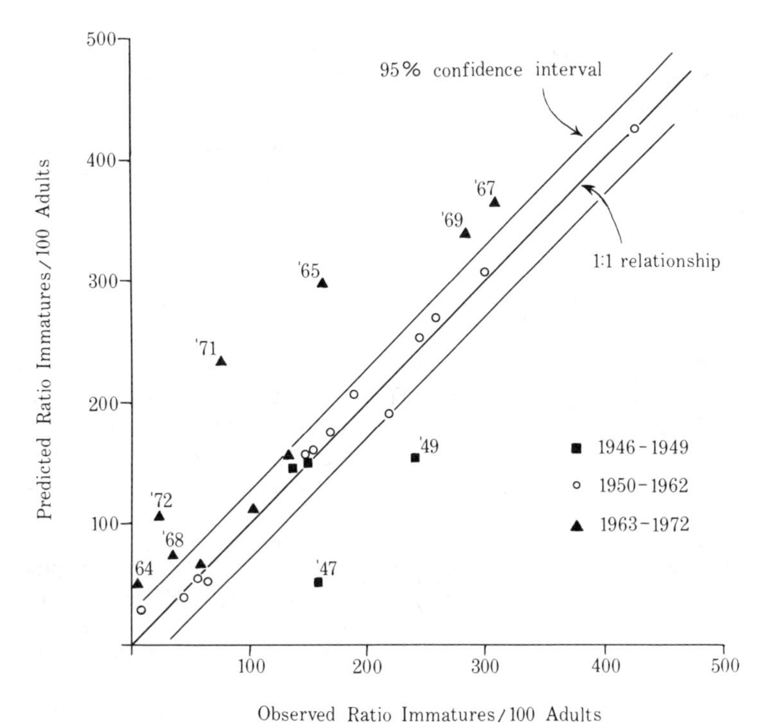

Figure 54. Predicted and observed age ratios of California Quail at Shandon, utilizing the predictive formula of Francis (1967). This graph depicts in another way the data shown in Figure 53.

departures have tended toward overestimation. Perhaps the increase in McMillan's quail population over the years has inversely influenced reproductive rate.

Despite these mathematical imperfections, there is clearly a positive relationship between rainfall and quail reproduction in the semi-arid hills of Shandon. There remains the matter of explaining the biological mechanism by which this relationship operates.

FORB YEARS VS. GRASS YEARS

Soil moisture can affect quail in only one logical way, and that is via nutrition. We must start with the precept, therefore, that the nature and vigor of vegetation which supplies food for the quail in some manner determines their reproductive drive.

McMillan (1964:707) was the first to note that years of high quail production were generally also years of exceptional growth of forbs and broad-leaved annuals. Specifically, he commented that in 1952—a banner quail year—a luxuriant and spectacular growth of annual vegetation was produced, predominantly of the species *Amsinckia douglasiana, Erodium cicutarium,* and *Lupinus bicolor*, the seeds of which are the most common quail foods of the region. In desert situations this would be known as a "wildflower year." Although at that time McMillan was dubious that nutrition actually regulated the quail hatch, he forcefully pointed out the optimum nature of the forb habitat for rearing young quail and for supporting them after they were grown. Subsequently, he has noted that in years of poor quail hatch the same hills are covered predominantly with stands of annual grasses rather than forbs.

Francis (1970) made a parallel observation in his quail pens at Berkeley. During the tenure of his study, he kept four pairs of quail in a 100 x 100 ft. pen where they had ample room to breed naturally. During the spring of 1963, the birds bred prodigiously, the four hens hatching a total of 52 chicks. That was the year that two of his penned hens double-brooded (Francis, 1965a). The following year, only 30 chicks were hatched and few of these survived. Of the vegetation in the pens in these two contrasting years, Francis states (p. 251):

> In 1963, a variety of annual forbs were present in abundance, the more prominent species in the spring including *Erodium cicutarium, Medicago apiculata, Medicago hispida, Geranium dissectum, Anagallis arvensis, Vicia sativa, Brassica campestris,* and *Ranunculus californicus*. Grasses, thistles, and other Compositae were also present.
>
> *Medicago* and *Erodium* are among important natural foods for quail . . . and presumably are highly nutritious; green leaves of *Vicia* were also extensively used by the quail in the enclosure.
>
> In 1964, forbs decreased greatly and grasses increased, in comparison with the previous year. None of the spring-blooming forbs were present, but thistles increased noticeably. Observation of feeding birds showed a greater use of grain supplements, compared to natural foods.

Grain was freely available to the birds both years, so the difference in reproduction was not a function of food per se but related in some way to the availability of forbs as food.

Campbell et al. (1973), in their excellent study of Scaled Quail in southeastern New Mexico, also noted a decided correlation between forb growth and quail reproduction. They state (p. 36): "It is unlikely that seasonal precipitation per se exerts an important influence on quail populations, but rather that precipitation controls some other factor which does exert an important or controlling influence. This factor probably is nutritional, most likely the food, vitamins, minerals, etc., produced by a diversity of forbs, all of which depend for their development and fruiting on adequate rainfall during definite seasons of the year."

I presume, therefore, that when quail have the opportunity to ingest the rich leaves of sprouting forbs, and the new seeds that follow fruiting, they breed more successfully than when this type of food is lacking or in short supply. The pattern of rainfall and temperature in winter and spring regulates the nature of the annual vegetation. Curiously, irrigation is no substitute for the natural regime of rainfall in producing the foods that quail need. In the vicinity of Shandon, coveys of quail living on the fringes of irrigated alfalfa have the same ups and downs of breeding success as coveys situated in the adjoining arid hills.

PRECONDITIONING OF BREEDING QUAIL

Another observation reported by McMillan and verified by both Francis and Erwin (Appendix C) is that, in years destined to produce good hatches of quail, the birds display a high level of aggressiveness and reproductive vigor during the breeding season, and this sexual drive carries through the summer. There is much fighting among the cocks, and persistent vocalization by those males that do not find mates. If a nest fails, the pair is highly likely to nest again. Conversely, in unproductive years sexual aggression is desultory, and when a nest fails, the birds give up trying, forming coveys and initiating the plumage molt as early as May or June. McMillan (1964: 710) summarized the point in these words: "This seems to indicate that habitat condition at the time of nesting is not the crucial factor regulating reproductive drive in quail. It appears instead that the birds are physiologically preconditioned by some factor, or combination of factors, presumably related to weather, operating during the winter or early spring prior to nesting."

Michael Erwin has elaborated this point with studies of quail behavior in 1972 (a poor nesting year) and 1973 (a highly productive year). His report offers many substantive details (see pp. 253–259, Appendix C).

If, therefore, my presumption is correct that qualitative nutrition obtained from fresh forb growth regulates quail reproduction, it would appear that the effect is imposed at an early stage in the nesting cycle. The die is cast prior to the hatching of the first eggs.

QUAIL NUTRITION AND REPRODUCTION

An early attempt to resolve this mystery was made by Lehmann (1953) in southern Texas. He noted that in exceptionally dry years the Bobwhite Quail on the King Ranch were desultory in breeding effort, and in fact the population of adults dropped to a very low level. In wet years the adults remained healthy and reproduction was successful.

Analysis of the Bobwhite livers for Vitamin A established a rough correlation between high reserves and good reproduction by the healthy adults. Conversely, in drought years (like 1951) the Vitamin A reserves reached a low level, adults became weak and heavily infested with parasites, and reproduction was minimal. As regards the effects of Vitamin A on breeding, Lehmann (1953:244) stated: "It did not appear that Vitamin A reserves were necessary for awakening the breeding urge. General breeding effort, however, was definitely associated with lush range and its frequent accompaniment of above-average feed. Quail, especially females, quickly attained high Vitamin A reserves after generous rainfall, and most successful breeding occurred after heavy May rains in both 1949 and 1950. With verdant range of short duration in 1951, survival of young was only 1.3 per adult as compared to 2.6 per adult in 1950, and 3.9 per adult in 1949.''

Vitamin A is obtained by quail from green food, but Lehmann's data showed conclusively that not all greens are high in this component. As a matter of fact, when his Bobwhites were dying of malnutrition in April 1951, and Vitamin A reserves were essentially exhausted, their food intake was made up largely of greens (57 percent). He points out that drought and heat are not conducive to the synthesis of carotene, which is the precursor of Vitamin A.

Hungerford (1964) followed Lehmann's lead by investigating Vitamin A levels in the livers of Gambel Quail in Arizona through three dry years (1954 to 1956) and one verdant year (1957). His data supported the general thesis that the year of good rainfall on the desert led to high reproduction of young, whereas the three preceding dry years were accompanied by meager reproduction. Likewise, Vitamin A storage in quail livers was high in the wet year and low in the dry years. He concluded (p. 141) that: ''Vitamin A or a closely associated substance derived from green plant material apparently acts as a stimulator which influences the rate of breeding in this desert quail.''

Fletcher (1971) undertook a controlled feeding experiment with California Quail to ascertain the effects of different levels of Vitamin A intake on the development and function of the reproductive organs. He found that birds on a diet completely devoid of Vitamin A ultimately used up their stores in the liver and showed typical deficiency symptoms including blindness, loss of weight, atrophy of the reproductive organs, and ultimately death. However, birds receiving some Vitamin A in the diet—from very light dosage to heavy—showed no consistent relationship between the amount of Vitamin A stored in the liver and reproductive vigor. His results could well explain the reproductive failure and mortality in a wild quail population that literally exhausts its Vitamin A supply, as observed by Lehmann in Texas Bobwhites, but would not seem to account for nonbreeding in desert populations of Gambel or California Quail that appear to remain in good health.

The observations of both Lehmann and Hungerford clearly demonstrate that lush green food in the diet is a concomitant of strong breeding in dry-land quails and that such a diet leads to high Vitamin A storage in the quails' livers. But neither their results nor Fletcher's experimental feeding study would appear to mark differential intake of Vitamin A as the primary factor regulating differential breeding from year to year. Francis (1970: 255) analyzed some quail livers during his study and found a substantially higher level of stored Vitamin A in birds taken in June, 1964 (a year of vir-

Figure 55. In semi-arid California, a "wildflower year" usually results in a good crop of young quail. Scene near Pinnacles National Monument. Jepson Herbarium photo.

tually complete reproductive failure) than in samples taken in the summer of 1963 (a year of good production). Apparently some other dietary component of lush green food is the critical ingredient, rather than Vitamin A.

GREEN FOOD AS A STIMULANT TO BREEDING IN DESERT RODENTS

Even as quail biologists have toiled to decipher the mechanism regulating breeding in arid lands, others have studied the same problem in desert rodents. F. S. Bodenheimer of Hebrew University in Jerusalem was a pioneer in proposing that some nutritive substance in green plants underlay the irregular breeding patterns of rodents in the Middle East. In 1946 he postulated that "increase of gonadotropic activity of voles in the field is due to . . . the ingestion of a plant gonadotropin which is subject to strong seasonal and cyclic fluctuations" (Bodenheimer and Sulman, 1946:256). In a general report on the natural history of the Merriam kangaroo rat in southern Arizona, Reynolds (1960:54) noted: "The two peaks of pregnancy in Merriam kangaroo rats corresponded closely with the periods of new vegetative growth of spring and late summer. . . . There is some evidence that nutrients contained in fresh vegetation have a stimulating effect upon breeding activity." Beatley (1969:723) studied reproductive cycles in desert rodents for 5 years on Jackass Flats in southern Nevada and made the following observation: "It is therefore concluded that the success or failure, and timing of rodent reproduction from year to year is dependent upon the presence or absence of winter annual . . . vegetation in the environment prior to or at the time of onset of the breeding season.'' Van de Graaff and Balda (1973:511) compared the breeding of kangaroo rats in two sites in Arizona, one of which received adequate rain to stimulate the growth of annuals while the other did not. They note: ''This significant difference in reproductive status of the two populations is strongly correlated with the amount of autumnal rainfall and the subsequent production of green vegetation.'' These and other similar reports seem to establish the fact that desert rodents, like California Quail, are triggered to breed by some nutritive stimulant ingested with the leaves of desert annuals, but what this component might be remains speculative.

THE POSSIBLE ROLE OF STEROIDS IN REGULATING BREEDING

In a continuing quest for an explanation of the linkage between nutrition and reproduction in quail, we turned our attention to the estrogenic steroids known to occur in varying amounts in the green leaves of some plants, particularly legumes. The dramatic discovery in Australia of the inhibitory effect of estrogenic substances in subterranean clover *(Trifolium subterranean)* on breeding in domestic sheep (Bennetts et al., 1946) opened a whole field of research in animal nutrition. Subsequently, Biely and Kitts (1964) and others have contributed to knowledge of the occurrence of estrogenic compounds in green leaves, and there is an extensive literature on the effects of these substances on ovarian activity in experimental mammals and birds.

Initially we collected samples of some of the forb leaves eaten in the spring by quail, and these were analyzed by Dr. John Oh of the University of California Field Station at Hopland, using the extraction procedure developed by Beck (1964) in Australia. Extracted samples were run through thin layer chromatography for quantitative analysis of phyto-estrogens (isoflavones and coumestrol) employing the Eastman chromatogram developing apparatus with silica gel incorporated with a flourescent indicator. In various samples, Oh identified four isoflavones—biochanin A, genistein, daidzein, and formononetin—plus coumestrol. We were uncertain whether these estrogenic hormones would stimulate or inhibit reproductive activity in quail, but feeding experiments conducted by Oh at Hopland and by Michael Erwin at Berkeley firmly established that sub-clover extract containing these substances inhibited ovarian development and egg production in penned female quail. Treated birds started egg laying a full two months after control birds on normal diet, and they produced few eggs (Table 10).

Erwin then initiated a program of collection of wild quail at Shandon to test the occurrence of phyto-estrogens in the crop contents of quail preceding and during the breeding season. Beginning in December, 1971, he collected by shooting approximately 10 quail a month through the 1972 and

TABLE 10.
Effect of diet on egg production in 3 pairs of California Quail

Diet	*Period of feeding trial*	*Onset of egg laying*	*Total egg production/pair*
Turkey starter*	3/10–6/30	April 8	62
Low energy Low protein†	3/10–6/30	May 8	28
Turkey starter** plus sub-clover extract	3/10–6/30	June 2	12

*Turkey starter contained 26% crude protein.
†Low protein diet contained 15% crude protein.
**Sub-clover extract contained isoflavones.

1973 breeding seasons, terminating in September, 1973. Crop samples for a given month were pooled, part of the material going to Bruce Browning of the California Department of Fish and Game Food Habits Laboratory for quantitative analysis of contents, and part to Dr. Oh for qualitative determination of the presence of phyto-estrogens. By great good fortune, 1972 turned out to be a very poor year for quail reproduction in the Shandon area (25 young/100 adults in the fall), while 1973 yielded a bumper crop of young quail (325 young/100 adults). We could thereby contrast the food intake and steroid consumption in years of very different reproductive success. Additionally, Erwin compiled a great deal of associated data on quail behavior and gonad development, reported in full in Appendix C of this volume.

Table 11 summarizes Browning's analysis of what the birds ate in 1972 and 1973. There was a notable increase in the consumption of green foods in the wet year (1973), as well as much higher representation of insects, which would mean a higher protein intake. The low consumption of greens in the dry year (1972) did not stem from unavailability, for there were forbs aplenty had the birds chosen to take them. Rather, the stunted forbs of 1972 seemingly were not highly palatable to quail, whereas the rank growing plants of 1973 were taken avidly.

Table 12 summarizes Oh's analysis of the occurrence of estrogenic compounds in the monthly samples of food consumed by the birds. Genistein and formononetin occurred consistently in medium to low amounts in the samples taken during the 1972 season, and biochanin A was present in a few of the samples. By contrast, these compounds were absent or virtually so from foods consumed by the quail during the 1973 breeding season, after the month of January. Coumestrol and diadzein were detected only periodically, and then largely in fall and winter samples. These results are reported in more detail separately (Leopold et al., 1976).

TABLE 11.
Volume percentage of green leafage, insects, and seeds in 86 California Quail crops taken near Shandon, 1972 and 1973 breeding seasons

Leafage	*Winter* 1971–72*	*Spring 1972*	*Summer 1972*	*Winter 1972–73*	*Spring 1973*	*Summer 1973*
Filaree (*Erodium*)	6.2	1.8	tr†	31.0	8.2	tr
Forbs (unidentified)	12.2	8.2	8.0	14.8	8.1	.2
Grass blades (*Graminae*)	8.3	tr	0	19.5	3.9	0
Clover (*Trifolium*)	tr	.8	.7	tr	17.5	0
Miner's lettuce (*Montia*)	0	0	0	0	7.1	0
Lotus (*Lotus*)	0	0	0	3.8	1.1	0
Lupine (*Lupinus*)	0	0	0	3.5	.3	0
Wild carrot (*Umbelliferae*)	0	0	0	0	3.2	0
Volume percentage of leafage	26.7	10.8	8.7	72.6	49.4	.2
Insects	tr	5.7	tr	tr	7.6	10.2
Seeds	73.3	83.5	91.3	27.4	43.0	89.6

*Winter = Dec., Jan., Feb. Subsequent seasons are 3-month intervals.
†tr = trace

We conclude that one of the mechanisms that may inhibit quail breeding in dry years is the presence of inhibitory compounds—largely formononetin and genistein—in green foods during years of stunted plant growth.

However, these preliminary findings are more suggestive than definitive, and a great deal of additional study will be required to fully understand the complexities of dietary regulation of quail breeding. Phytoestrogens may play a role in suppressing reproduction in a dry year, but what is the stimulus that generates abnormal reproductive vigor in a wet year? There are innumerable other chemical compounds produced by growing plants that may affect quail fecundity positively or negatively. Moss et al. (1974) are studying the same problem in Scottish Red Grouse, and they also conclude that qualitative nutrition of the female grouse determines her success in rearing chicks.

LEGUME SEEDS AS A POSSIBLE STIMULUS TO BREEDING

Erwin (1975 and Appendix C) expounds the possibility that it is not the green leaves but rather the seeds produced by early flowering desert forbs that constitute the stimulus to sustained quail breeding in a wet year. In particular, he emphasizes the importance of legume seeds of the genera *Lupinus, Lotus,* and *Trifolium* as dominant foods in the diet of his study population in the wet year of good reproduction (1973) as compared to the previous year of poor reproduction (1972). In 1973, legume seeds constituted 29.8% of the summer diet, in contrast to 11.6% in 1972. Grass seed of lesser nutritive value made up the difference in 1972. Erwin is inclined to discount the green component of the diet as the stimulant to strong breeding effort. Although he acknowledges that phytoestrogens in stunted greens may play a part in discouraging breeding in a dry year, he points out correctly that there still must be some *stimulatory* component of the diet in a wet year. During incubation a hen quail forages for only a few hours a day. If in that time she finds an abundance of new seeds, her strength and reproductive drive may be sustained, both physiologically and psychologically. Although this hypothesis does not in my opinion fit all the available facts, it certainly warrants serious consideration. For a full discussion of Erwin's view, see Appendix C.

In any event, the riddle of quail nutrition as it bears on quail reproduction is not resolved, and I foresee long and tedious years of study ahead before this complex relationship is fully understood.

RAINFALL AS A DIRECT STIMULANT TO BREEDING

There is a possibility that rainfall per se has a sexually stimulating effect on quail which does not depend on a link with nutrition. In the deserts of Australia, a number of bird species are known to react directly to the psychological stimulus of falling rain, producing eggs within a few days after rain falls irrespective of the season of year. This interesting situation is discussed at length by Serventy (1971) in his review paper on "Biology of Desert Birds." He finds that the astonishing adaptability of the Australian avifauna to sporadic rainfall is not duplicated in the desert areas of any other continent. The birds of desert regions in Africa, Asia, and America all tend to breed seasonally under control of a light regime, although a breeding season might be suppressed or missed completely in a dry year. Even among the Australian birds often cited as responding directly to rainfall, there have been found to exist nutritional links which affect sexual development. Frith (1959, 1967) established that Gray Teal and some other species of waterfowl are stimulated to instant sexual activity by rising water levels in the billabongs. Water level is, of course, a function of rainfall and runoff. He noted that rising water induced the development of an abundant supply of insects suitable for the nourishment of the ducklings, but the basic trigger that induced breeding behavior was thought to be largely psychological. Subsequently, however, Brathwaite (in press) and Brathwaite and Frith (1969) have concluded that the sudden availability of nutritive food might itself be part of the mechanism that initiates breeding, irrespective of changes in water level; so the solution in the case of some of these species may come back to nutrition, after all.

The mass of evidence concerning the breeding of the California Quail on arid or semi-arid ranges points to regulation by nutritional cues, as we have postulated.

TABLE 12.

Estrogenic isoflavone content of pooled quail crops (both sexes) taken near Shandon during a winter and spring leading to poor reproduction (1971–72) and the following year leading to abnormally high reproduction (1972–73)

Collection date	Number pooled crops	Levels of phyto-estrogens (High, Medium, Low)					Breeding success
		Biochanin A	*Coumestrol*	*Daidzein*	*Formononetin*	*Genistein*	
1971							
Dec. 12	4	L	—	—	—	L	Poor 25 imm./ 100 ad.
Dec. 13	5	—	—	—	—	M	
Dec. 28–31	7	—	—	—	L	L	
1972							
Jan. 17	4	L	—	—	L	L	
Jan. 22–24	12	L	—	—	L	L	
Jan. 28–30	18	—	—	—	M	L	
Mar. 4–5	5	—	—	—	M	M	
Mar. 27–30	4	—	—	—	L	L	
Apr. 16	3	—	—	—	L	L	
May 1	3	L	—	—	L	L	
May 13–14	6	—	—	—	L	—	
June 3–4	3	—	L	—	L	L	
July	6	—	—	—	—	—	
August	4	—	—	—	—	—	

December	7	—	—	—	—	M	
1973							
Jan. 3–5	9	—	—	—	—	H	
Jan. 5–27	10	—	—	—	—	L	
Mar. 30	7	—	—	—	—	—	Excellent 325 imm./ 100 ad.
April	2	—	—	—	L	—	
June	5	—	H	—	—	—	
July	6	—	—	—	—	—	
August	3	—	—	—	—	—	
September	6	—	L	—	—	—	

CONCOMITANT BENEFITS OF RAINFALL TO QUAIL REPRODUCTION

My discussion up to this point has centered on the factors which stimulate or inhibit breeding behavior in adult quail on arid ranges during the spring and summer period of nesting. However, there are important additional considerations concerning the survival of young, once they are produced.

Foremost of these is the supply of easily available food required to support newly-hatched chicks through the juvenile stages of life. A wet season with generous forb growth would clearly produce an abundance of insects, which constitute a small part (but perhaps a crucial part) of the diet of very young chicks. More obviously, however, the supply of new seeds dropping on the surface of the ground from the maturing forbs would be available for the major sustenance of the young birds. I suspect that inexperienced young quail might have difficulty scratching a living from old seeds buried in the soil or in the duff layer. The same forb growth that in one way or another encourages adult quail to breed also supplies the sustenance to support the young birds, not only during their period of development but on through the subsequent winter.

Another corollary benefit of a wet year would be the added sources of drinking water. Whereas adult quail in desert situations may subsist comfortably on succulent greens and metabolic water produced from digestion of carbohydrates (seeds), it is probably not often that young quail survive the first summer of life without access to free drinking water. Sumner (1935) reared a brood of quail chicks on succulence alone in the cool, moist hills near Santa Cruz, but it does not follow that quail in desert situations would fare so well. In any event, we know that most quail broods come to water more or less regularly, and I presume that drinking favors their welfare.

RELATION OF POPULATION DENSITY TO BREEDING SUCCESS

There is a general maxim in the field of wildlife biology that reproductive success in a population of wild animals is often inversely proportional to the density of the population. To what extent may quail density affect reproduction, irrespective of rainfall and conditions of the habitat?

In our observations of the quail population at Shandon, the effect of density is minimal, or at least is largely obscured by the more immediate relationship of nutrition of the adults and resources for the rearing of young. Through intensive management, McMillan has built up the population on his own ranch to a level of a bird per acre in even the poorest year, and much higher in years of good hatch. The productivity of this population, as measured by age ratios taken in fall samples, is neither higher nor lower than that of nearby populations existing at much lower densities. In a year of good rainfall and forb growth, the proportion of young in McMillan's birds is essentially as high as that of other quail flocks in the region. Conversely, in a poor year, reproduction is equally low for all quail around Shandon, whatever the number of breeders per acre. In short, the productivity of arid land quail appears to be more or less *density-independent*. Biologically, this is an interesting observation, in that it appears to contradict a widely accepted generalization of population dynamics.

In more humid and northerly ranges, however, the conventional concept of "inversity" seems to be valid. One of the most careful and searching studies of population dynamics in the California Quail was conducted on southern Vancouver Island by Barclay and Bergerud (1975). They found that quail reproduced each year but that chick production and survival varied inversely with the density of the breeding population. They state (p. 320): "The fall juvenile:adult ratios reflected the observed increase between years, while the adult winter and summer mortality appeared to remain relatively constant. Thus the population trends during this study appear to be largely a result of differential recruitment of chicks rather than compensatory mortality of adults at any time during the year. Furthermore, the recruitment of chicks was inversely correlated with the spring density." No clear correlation was evident between quail productivity and weather parameters preceding or during the breeding season. But a severe winter in 1968–69 led to a marked decrease in numbers, followed by compensatory increases in 1969 and 1970, restoring the population to its "normal" level. Moreover, they found some behavioral differences in chicks implying genetic differences in the "quality of the stock" produced in years of high versus low breeding populations. This intriguing suggestion calls for further study.

To summarize, in the more northerly and humid ranges of the California Quail the birds breed successfully every year, but the actual production of young per pair is highest when the level of the breeding population is lowest. Conversely, in the more arid southerly ranges, winter and spring rainfall is so important in regulating breeding that influences of population density are largely obscured.

10
QUAIL MORTALITY

MORTALITY VS. NATALITY

In any stable unit of quail range, the number of quail that die will exactly equal the number that are hatched. That is to say, over a period of years in which the mean level of a population remains about the same, there can be no excess of deaths over births, or vice versa. As noted in Chapter 8, the average "turnover" in a population of California Quail will vary locally from 59 to 77 percent, meaning that of the grown birds in an autumn population, that proportion will die and be replaced from one year to the next. The actual number of births and deaths is much greater than that, since substantial numbers of chicks hatched in the spring never reach adulthood and hence are not counted in turnover statistics.

A wild species like the California Quail has evolved a high rate of productivity to compensate for inevitable high losses from predators and other causes. Population continuity is achieved by the mechanism of overproduction (Leopold, 1954). To take a specific example, let us say that in a given watershed there are 100 quail in the spring. These birds pair and nest, and by late summer they have raised a crop of young. Some of the nests were destroyed by ground predators like skunks, raccoons, and feral house cats, forcing these pairs to re-nest and try again. A good many of the chicks and even some of the adults, especially incubating females, likewise are caught by predators. Nevertheless, following a year of good breeding, the whole population may have increased to perhaps 250 birds. If the age ratio of the population is sampled by hunting, it will be found to consist of roughly 65% young and 35% adults. From this point until the following breeding season, the survivors are harassed by predators and other dangers, and many additional birds are killed. Come spring again, the chances are that the population will be back down to around 100 breeders. The cycle then repeats itself.

The 150 birds that were lost may be looked upon as "biological surplus," and many recent studies suggest that these individuals are predestined to perish in one way or another because the habitat will not support them. Each unit of game range, like the watershed in the example above, has an arrangement of cover, food, and water that will tend to maintain just so much game, year in and year out. It is part of the natural scheme of things that the breeding stock will produce a crop of young considerably in excess of what is needed to replace losses among the adults. The excess supplies hunting and likewise feeds the predators. By the same token, a forest drops infinitely more seed than is needed to replace the dead trees. The extra seed feeds chipmunks and small birds. A pair of bluegills may produce thousands of young that feed the pike and bass, only two having to survive to replace their parents.

One frequently noted aspect of predation is that it is most effective when surpluses are large. Predator kills decrease in frequency as the population of the prey shrinks, meaning that predation is *density-dependent*. Thus, 250 quail in our arbitrarily defined habitat may furnish many good meals for hawks, owls, and other flesh eaters, but as the population drops toward 100 the rate of predation falls off until finally it virtually ceases. A point of diminishing returns is reached where it is not worth the predator's time to try to seek any more quail, the alert and experienced survivors being ensconced in the most "secure" parts of the habitat. This point is determined by the nature and distribution of the cover and its proximity to food and water. In the parlance of wildlife biology, the area may be said to have a "carrying capacity" of 100 quail.

If the following spring is dry and few young are produced, losses to predation will be very low, and most of the 100 carry-over adults will live through another year to await the next breeding season. The probability of life for an adult quail, therefore, will be inversely proportional to the number of young it succeeds in rearing. In an actuarial sense, the individual quail's worst enemy is not the predator but its own progeny. As noted in Chapter 8, high reproduction begets high mortality, not just among the young but among the parents as well.

Hunting and predation are not the only sources of mortality. Some birds may succumb to disease or parasites. In areas where agricultural poisons are widely distributed, quail may be debilitated or killed outright by ingestion of poison bait or of insects killed by spray. Other birds suffer accidental death by flying into wires or moving automobiles. Some of these sources of mortality are preventable, others are not. This chapter will consider the more important sources of mortality in the California Quail.

QUAIL PREDATORS

Predators and predation account for most of the mortality inflicted on populations of California Quail. However, there is surprisingly little quantitative information on the precise numbers of birds taken by individual predators. Even the most intensive studies have not yielded actuarial details on losses. Perhaps one reason for this dearth of information is that for the most part the range of the California Quail is not covered with snow in which the record of mortality can be read. Errington (1934) followed the tracks of Bobwhite coveys in Wisconsin and Iowa and recorded with considerable precision the instances of predation and what predator was involved. In the absence of continuous snow, that type of record is impossible to obtain. Most of the records summarized here are based on subjective observation and inference.

Cooper Hawk. All observers agree that the Cooper Hawk is the primary natural enemy of the California Quail. It flies swiftly along the ground or darts through trees and brush, pouncing upon any small prey that is sur-

Figure 56. The Cooper Hawk is the most efficient and persistent predator of California Quail. C. W. Schwartz photo.

prised away from dense cover. Every autumn, migratory Cooper Hawks descend to the foothill quail ranges from their breeding grounds in coniferous forests, and from the moment of their arrival until departure in early spring the quail are given no respite. Genelly (pers. comm.) documented many attacks of Cooper Hawks on the quail that he had under study in the Berkeley Hills. Several times he had to abandon his quail trapping program because the Cooper Hawks would kill the quail caught in his traps. Each year McMillan observes the continuing harassment of quail around his homestead by transient Cooper Hawks. In the period 1959 to 1961, 18 Cooper Hawks were taken in the vicinity of the McMillan ranch for a food habits study being conducted by E. W. Jameson, Jr., of the Davis campus; two contained remains of quail. Michael Erwin, during his quail studies near Shandon, observed repeated Cooper Hawk attacks, and he found the remains of 11 quail killed by this predator at a single waterhole on the Camatta Ranch. On the desert fringe of the San Bernardino National Forest, Rahm (1938) speaks of four kills made where quail were congregated near watering troughs. In the Santa Cruz Mountains, Sumner (1935:325) characterizes this hawk as the "arch-enemy of quail," but states: "Of all the tries which I have personally seen Cooper Hawks make for quail, none happened to be successful, but this was in a region where cover is exceptionally abundant." At the Dune Lakes Club in San Luis Obispo County, Glading et al. (1945) examined the stomachs of 25 Cooper Hawks of which 3 contained quail; the high quail population on this property was intensively managed, exceeding 4 birds per acre. Grinnell et al. (1918:532) recorded shooting a Cooper Hawk in the act of eating a quail. Fitch, Glading, and House (1946) observed the rearing of two broods of Cooper Hawks on the San Joaquin Experimental Range in Madera County; of 41 prey items brought to the nests, 7 were California Quail.

In a quantitative sense, the actual number of recorded quail kills by Cooper Hawks is an inadequate measure of the impact of this predator on the quail population. Disturbance and harassment are perhaps more serious than direct mortality. The mere presence of Cooper Hawks renders a great deal of quail range with marginal cover unusable during the fall, winter, and spring.

Sharp-shinned Hawk. The sharp-shin is the small counterpart of the Cooper Hawk, but it is less abundant in the range of the California Quail. Sharp-shinned Hawks undoubtedly take quail on occasion. Glading et al. (1945) found quail in 2 of 6 hawks taken at Dune Lakes. Hunting near Auburn, I crippled a quail which, crossing a valley in labored flight, was overtaken by a sharp-shin, driven to the ground, and killed. De Fremery (1930) describes the "freezing" reaction of a covey of quail in the presence of a Sharp-shinned Hawk. McMillan saw a sharp-shin roll down a bank with a quail. The quail was partially defeathered but succeeded in escaping.

Marsh Hawk. Marsh Hawks tend to frequent open fields and grasslands rather than the brushy habitat of quail. Where their ranges overlap, however, some predation may ensue. On the Dune Lakes Club, Glading et al. (1945:178) observed: "The marsh hawk . . . preyed quite freely on fledgling quail during the nesting season. At marsh hawk nests located in the heart of the quail concentration, more than 20 percent of all items brought in were young valley quail." Two of 6 stomachs of adult birds contained remains of young quail.

Horned owl. This nocturnal predator will attack quail that it finds roosting in inadequate shelter. In addition to the occasional bird that it catches, the disturbance constitutes a danger by sending quail flying into the night where they may be hurt in collision or be exposed to attack by other predators. Where roosting cover is adequate, the Horned Owl poses little threat to quail.

Bobcat. Grinnell et al. (1918:532) state that "Wildcats are about the worst enemies of these birds [quail]." Subsequent studies have produced little evidence to support this assertion. Sumner (1935:329) comments: "Of 156 bobcats taken by state trappers on nine California game refuges, only 3 percent of the stomachs contained quail (McLean, 1934), while out of eight additional stomachs of bobcats taken in southern California from state-operated quail refuges . . . none contained quail. Dixon (1925:36) found that out of 218 bobcat stomachs, all kinds of birds together amounted only to about 4 percent of the total food. Bobcats live chiefly on rabbits, wood rats, gophers and other rodent competitors of quail. . . ." At Dune Lakes Club, Glading et al. (1945) found quail remains in one bobcat stomach out of five examined. Leach and Frasier (1953) analyzed the contents of 53 stomachs taken in the immediate vicinity of quail watering devices, where the birds would appear to be highly vulnerable to predation. Yet only one of the bobcats contained the remains of a quail. Bobcats, like Cooper Hawks, certainly harass quail, both during the day and on the roosts at night. They may force the abandonment of inadequate coverts. But there is little indication that they catch many birds.

Other predators. Unimportant predators known to catch an occasional quail are the Red-tailed Hawk, Prairie Falcon, Sparrow Hawk, Coyote, and Long-tailed Weasel. The house cat harasses quail and may drive them from the vicinity of a yard or a feeding station (Sangler, 1931), but there is little evidence that they catch many quail in wild situations. Hubbs (1951) analyzed the stomach contents of 219 feral cats taken in the Sacramento Valley and recorded one California Quail. Feral cats, like bobcats, prey mostly on rodents.

Nest predators. Predation on quail nests is much more easily documented than predation on the birds themselves. Studies of many species of ground-nesting game birds have shown that half or more of the nests will be broken up by predators, and the California Quail is no exception. Reference already has been made to the quail nesting study conducted by Glading

(1938a) on the San Joaquin Experimental Range (see Figure 34, Chap. 6). Of 93 nests, 30 were broken up by Beechey ground squirrels, 6 by house cats, 6 by coyotes, 5 by skunks, 3 by bobcats, 3 by California jays, and 1 by a gray fox. This rate of attrition is probably above normal because of the abundance of ground squirrels, but the results do emphasize the role of mammalian predators in nest interference. Additional animals that may eat quail eggs in the nest are raccoons, crows, magpies, and several kinds of snakes (Sumner, 1935:328; Twining, 1939:33). In some situations, red fire ants *(Solenopsis)* may enter pipped eggs and consume the helpless chicks (Emlen and Glading, 1945:45). Some nests are trampled by domestic livestock or broken up by farm machinery.

The point should be reiterated, however, that the California Quail evolved in the presence of all these native predators and can accommodate considerable loss of eggs and nests without seriously depleting the population. In a good breeding year, pairs that lose one nest will try again, or even a third time, in the effort to bring off a brood. In a year of scant breeding, the loss of first nests is more serious since there is little renesting. The best insurance against undue nest predation is the maintenance of adequate nesting cover, so that nests are widely scattered and difficult for the predator to find.

DISEASE AND PARASITES

Like any other native vertebrate, the California Quail harbors various diseases and parasites. Only two diseases are of any consequence, and these appear to be sporadic and of minor importance in regulating population levels in the wild.

1. *Haemoproteus lophortyx,* a blood parasite allied to malaria, was first noted in quail by O'Roke (1928). Subsequently the organism was found to occur quite commonly in wild California Quail taken at the San Joaquin Experimental Range (Herman and Glading, 1942), at Dune Lakes Club, San Luis Obispo County, and near Bitterwater in San Benito County (Herman and Bischoff, 1949). Frequency of infection varied from 45 percent to 84 percent, but most cases seemed to be benign without obvious signs of debility. The *Haemoproteus* organism is transmitted by bloodsucking hippoboscid flies that live in the plumage of the quail (Tarshis, 1955).
2. *Coccidiosis* is an intestinal disease of quail caused by organisms of the genus *Eimeria,* one of the protozoan genera identified by Lewin (1963) as normally present in the intestinal tracts of wild quail. Liburd (1969) identified coccidia in 73 percent of 137 quail collected in the Okanagan Valley of British Columbia, and Liburd and Mahrt (1970) described two new species of *Eimeria* from this sample. Under circumstances of stress, particularly in penned birds, coccidia may proliferate and

cause considerable mortality. Transmission is through droppings. Herman and Jankiewicz (1942) confined 15 quail suffering from *coccidiosis* in a pen with wire floor, so that droppings were out of reach, and the incidence of infection rapidly dropped. Within a week, very few oocysts of *Eimeria* could be found in the droppings, indicating that the "larger number of parasites likely to injure the health of the birds is maintained only with continual reinfection" (p. 149). In the wild, droppings are scattered, and pathologic reinfection is unlikely except in areas of great concentration of quail.

3. *Other internal parasites* that occur in wild California Quail include various roundworms and tape worms (O'Roke, 1928; Chandler, 1970), the gapeworm (Herman, 1945), and the quail heartworm (Weinmann et al., in press). There is no evidence of mortality in wild populations from worms, although the gapeworm may constitute a problem in penned flocks.
4. *Hippoboscid* flies of two species (*Stilbometopa impressa* and *Lynchia hirsuta*) live in the plumage of wild quail, but they are not common. I have occasionally seen the larger fly (*Stilbometopa*)—about the size of a housefly—crawling through the feathers of shot quail. The smaller and narrower species (*Lynchia*) was found by Sumner (1935: 244) on 20 percent of the birds examined, but unless specifically searched for it is rarely observed. Hippoboscids themselves have little direct effect on the health of quail, but they are significant as vectors of the *Haemoproteus* blood parasite.

All in all, parasites and disease are not an important source of mortality in wild populations of California Quail. Sumner (1935:244) noted the occurrence of some unidentified disease among quail on the campus of Mills College, Oakland, but the outbreak soon subsided. We have no further reports of die-off from any part of the quail range.

AGRICULTURAL POISONS

Quail mortality stemming from ingestion of agricultural poisons has in the past constituted a significant source of loss in quail populations. Occasional losses still occur. With the intensification of agriculture in California, increasing amounts of chemical compounds are being applied to the countryside for the following purposes: (1) to kill rodents; (2) to kill insects; and (3) to kill weeds. Direct mortality of quail derives from the use of some rodenticides and insecticides. Herbicides used in weed control have an indirect effect on quail by removing food and cover plants.

Rodenticides. Among the early reports of quail poisoning was the account by Linsdale (1931) of substantial losses following the distribution of grain treated with thallium for ground squirrel control. Linsdale compiled reports of wildlife deaths in various parts of California and listed 713 dead quail that had been found in thallium treated areas. He states (p. 103): "It has been definitely established that quail will eat thallium-treated barley, and that they are killed by it. Also, the facts that whole coveys have disappeared from poisoned ground and that the species became so reduced in whole counties as to be not worth hunting have been observed.'' Shaw (1932) reports that approximately 8 kernels of treated barley constitutes a lethal dose for a quail, which figures to 12 mg/kg body weight. The widespread use of thallium was discontinued in the 1940's with the introduction of other rodenticides.

Strychnine was widely used in the past as a rodent poison, and it is mildly toxic to quail, though highly toxic to some other birds. Emlen and Glading (1945) fed strychnine-coated squirrel bait to quail over extended periods without actually killing any of their experimental birds. Today, strychnine is still in use for blackbird control, but it is not a threat to quail.

Sodium fluoroacetate (compound 1080) came into general use as a rodenticide in the late 1940's. Treated grain applied either from airplanes or from the ground has proven to be a highly effective way to control the Beechey ground squirrel, and populations are held at very low levels by distribution of bait every two to three years. This compound is highly toxic to a wide spectrum of animals, including quail. Its application in California is under the jurisdiction of County Agricultural Commissioners and the State Department of Agriculture. Enormous amounts of 1080-treated grain are distributed each year, with possible risk to wildlife other than the target species. Table 13 shows by counties the quantity of 1080-treated grain put out in California in the years 1969 and 1970. The average application in recent years has been over 400,000 pounds, most of it scattered in grazing lands, which constitute the primary quail range. Sayama and Brunetti (1952) tested the toxicity of 1080 to captive quail and concluded (p. 298): ''Sodium fluoroacetate is extremely toxic to California quail. Although insufficient birds were available to make an accurate determination of the minimum lethal dose, it was found to lie between 1 and 5 mg/kg of body weight.'' Although treated grain is dyed bright yellow to discourage consumption by birds, and formulations as presently mixed are low in toxic material (1.5 oz. per 100 lbs. of grain, as compared with 5 to 10 oz. used in earlier years), the fact remains that quail are regularly exposed to this hazard where poison is distributed. The actual losses of quail to 1080 grain are impossible to measure because the aerial distribution of the bait is so diffuse over extensive areas of rough terrain. Many other species are secondarily affected, including especially coyotes that scavenge on the dead rodents. In the vicinity of Shandon, coyotes are rare as a result of periodic distribution of 1080 squirrel bait.

Insecticides. For a period of two decades, from the late 1940's into the 1960's, dangerous insecticides were distributed on agricultural lands, and to a lesser extent on rangelands, often with little knowledge or concern about deleterious effects on wildlife. Under some circumstances, substantial

TABLE 13.
1080 use by the California counties for rodent control in fiscal years 1969 and 1970 (Data from *Weed and Vertebrate Pest Control Reports,* Calif. Dept. Agriculture, Sacramento)

	Pounds of 1080-treated bait used			
	Economic rodent control		*Plague control*	
County	*1969*	*1970*	*1969*	*1970*
Alameda	8,509	7,851	80	157
Calaveras	475	317		
Colusa		540		
Contra Costa	1,720	1,373		
Fresno	73,300	301		
Glenn	5			
Humboldt		1,949.5		
Kern { aerial	33,750	120,085	2,000	4,700
Kern { ground	10,061	16,058	3,280	
Kings	2,545	15,431		
Lassen	171	450		
Los Angeles	1,035			
Madera	6,574	8,074		
Mendocino	801	1,334		
Merced	14,220.5	26,266		
Modoc	24,835	18,160		
Monterey	18,655	43,644	1,418	930
Orange	3,841	1,340		
Riverside	8,783.5	9,084.5		
Sacramento	382	1,906		
San Benito	17,199	16,878		
San Bernardino	2,085	1,480		
San Joaquin	4,499	4,570		
San Luis Obispo	83,196.25	78,824	75	67
Santa Barbara	4,239.25	3,477	333	325
Santa Clara	10,522		2,882	821
Santa Cruz	23.5	28.5	19	11
Siskiyou	3,440	1,120		
Solano	137	275		
Stanislaus	16,040	22,791.5		
Sutter	147			
Tulare	1,925	48,598		
Ventura		5,222	16,235	5,222
Yuba	125	78		
Total	353,241	457,506	26,322	12,233

numbers of wild birds, mammals, and fish were inadvertently killed or debilitated. Perhaps the best-known case concerned the effects of DDT on hawks and some water birds. This persistent chlorinated hydrocarbon is concentrated in food chains, and when ingested by predatory birds or fish-eating birds accumulates in their bodies and has the effect of inhibiting the normal deposition of egg shell, through interference with enzyme functions. The Duck Hawk or Peregrine Falcon has been brought close to the point of extinction in North America and Europe by DDT interference with reproduction (Hickey, 1969). The Brown Pelican on the California coast essentially ceased to breed in the late 1960's for the same reason (Gress, 1970), and it was not until the general use of DDT was banned by the Environmental Protection Agency in 1972 that successful breeding resumed. Rosene (1969:209) cites evidence of substantial loss of Bobwhites in South Carolina following dusting of soy bean fields with DDT. Recurrent instances of wildlife poisoning by insecticides led to numerous investigations and ultimately to the adoption by state and federal agencies of rigid regulations over the use of insecticides.

Quail and other gallinaceous birds are highly susceptible to poisoning by some other chemical insecticides. In the Southeast, the aerial application of bran treated with heptachlor and/or dieldrin to poison the introduced fire ant killed out Bobwhite populations over large areas. Clauson (1959) reported complete extermination of Bobwhites in test areas treated with these chemicals at the rate of 2 lbs/acre (technical material). Post (1951) observed severe effects on Chukars, Pheasants, and Sage Grouse from treatment of millions of acres of Wyoming rangelands with toxaphene and chlordane to control grasshoppers. No comparable field studies have been made of effects of insecticides on California Quail, but Rudd and Genelly (1956) compiled a complete dossier of all the pesticides then in use and categorized their toxicity to wildlife, including in many instances to California Quail. The following discussion of possible impact of insecticides on the California Quail is based on their compendium and on the advice of Dr. W. W. Middlekauff of the Department of Entomology, University of California, Berkeley.

On California rangelands, the two insecticidal programs that potentially might have impact on California Quail are aerial spraying for (1) grasshopper control and (2) beet leafhopper control. Both involve application of chemicals to rangelands, where most of the quail live.

Under current practice, grasshopper outbreaks are checked by one of four types of application: (1) malathion applied as a concentrated liquid spray; (2) diazinon applied as a spray; (3) carbaryl (organic carbamate) also applied as a spray; and (4) toxaphene-treated bait (bran or rolled oats) applied in the immediate area of a hatch of young grasshoppers. The first three chemicals are of low toxicity to wildlife, including quail, and pose relatively little danger. Toxaphene is rated as "moderately toxic" to birds, but this method of grasshopper control is not widely used in California.

None of these chemicals is persistent in the environment; hence there is no risk of cumulative effects.

Beet leafhoppers breed in foothill rangelands, feeding on Russian thistle and other winter annuals. As they mature they migrate down canyons toward irrigated bottomlands where sugar beets, tomatoes, and other truck crops are grown. Leafhoppers transmit a serious virus disease of beets ("curly top") and of some other crops. Control is achieved by spraying rangelands where the leafhoppers breed. In the past, DDT was sprayed on breeding areas, but this involved considerable risk to quail and other wildlife. With the current ban on DDT, malathion has been substituted, and, according to Rudd and Genelly (p. 108), malathion is of low toxicity to wildlife; no kills of either birds or mammals have been reported.

In other words, the regulations governing use of insecticides on rangelands have forced the abandonment of the more dangerous and persistent chlorinated hydrocarbons like DDT, and shifted the control measures to less persistent and less hazardous materials. The probability of large-scale losses of quail will certainly be less in the future than in the past.

On irrigated croplands, a wide variety of insecticides is in use, and quail living in coverts adjoining these agricultural areas might be exposed to some very toxic chemicals. This is particularly true of quail habitat adjoining orchards, vineyards, and cotton fields. The shrinkage of quail populations stemming from clearing large areas for intensive cropping may at times be extended to bordering areas by the drift of insecticidal sprays.

Herbicides. There is no evidence that the commonly used herbicides are toxic to gallinaceous birds such as the quail. Rudd and Genelly (1956) summarize feeding tests made with a variety of weed poisons, including 2,4-D, 2,4,5-T, "dalapon," sodium chlorate, and others, leading to the conclusion that none of these is likely to affect quail directly.

On the other hand, the widespread use of herbicides to control weeds in cultivated fields, or to kill roadside vegetation, has an enormous effect on the quail habitat. The broad-leaved forbs that are the target of much herbicidal spraying are the very plants that supply much of the quail diet. Herbicides applied from an airplane to control weeds in grain fields often drift across adjoining pasture lands, selectively removing lupines, fiddle-neck, clovers, and filaree, thereby favoring the competing grasses. Roadside cover, essential to quail in some regions of California, is systematically killed to keep the road-shoulders "clean" and to reduce the breeding habitat for leafhoppers. The common Russian thistle or tumbleweed, for example, is listed as a noxious weed and is subject to spraying and later burning by road maintenance crews. Tumbleweed not only serves as good quail cover, but it is an important pioneer plant in stabilizing denuded and eroding soils exposed by cultivation, overgrazing, or road grading. Both the landscape and the quail would profit by preserving tumbleweed and other roadside vegetation in areas where the danger to crops from leafhoppers is low.

The overall effect of herbicide usage on the habitat of the California Quail may be more deleterious to the species than the application of rodenticides or insecticides that kill birds directly.

STARVATION

California Quail are subject to starvation mainly in those parts of the range that are occasionally covered with deep snow or glazed with ice. Such conditions occur from time to time in the more northerly and easterly habitats, such as in Washington, Oregon, northern and eastern California, Nevada, and Utah. Snow or ice can render normal food supplies completely inaccessible to the birds, leading to loss of weight and loss of resistance to low temperatures. Nielson (1952) reports a major die-off of California Quail in the Uintah Basin, Utah, in 1948–49—a winter of deep snow and sub-zero temperatures. A new breeding stock had to be imported to restore the population. William Molini (pers. comm.) of the Nevada Department of Fish and Game tells me that restocking of quail coverts depleted by hard winters is a regular management procedure in Nevada.

Throughout most of California and all of Baja California, severe winter conditions rarely if ever occur. At the same time, there may be definite shortages of food in years when seed crops do not mature or when food plants are eliminated by herbicides or clean cultivation. In these circumstances, the birds are forced to forage widely, often far from protective cover, to find nourishment. Mortality increases, if not by outright starvation then by predation or other hazards associated with extensive travel.

HUNTING AS A SOURCE OF MORTALITY

The rationale of hunting is that it removes part of the autumn surplus of birds that are "expendable" and doomed to die from one cause or another. Presumably, some proportion of a population can safely be removed by hunting without prejudicing the breeding stock for the following year.

The subject of regulating the hunting take will be discussed in detail in Chapter 14. It will be brought out that an average kill of 30 to 40 percent of a healthy autumn quail population is a reasonable harvest goal. Under careful management, the kill may even exceed this level.

In the more arid portions of the California Quail range, however, it must be remembered that few years are "average." In a dry year of poor reproduction, there is relatively little latitude for trimming the autumn population, and the hunting removal should be minimal. This happens almost automatically because the reduced number of birds, occurring in small coveys made up largely of experienced adults, makes hunting difficult and generally unproductive. Conversely, in a wet year of high production, a large surplus of young birds can withstand a heavy harvest—perhaps up to 40 percent or more.

In any large block of quail range, such as a county, only a portion of the quail habitat will be subjected to hunting. Some areas are inaccessible.

Figure 57. Even in the arid ranges of southern California, snow occasionally imposes periods of stress on the quail. M. Erwin photo.

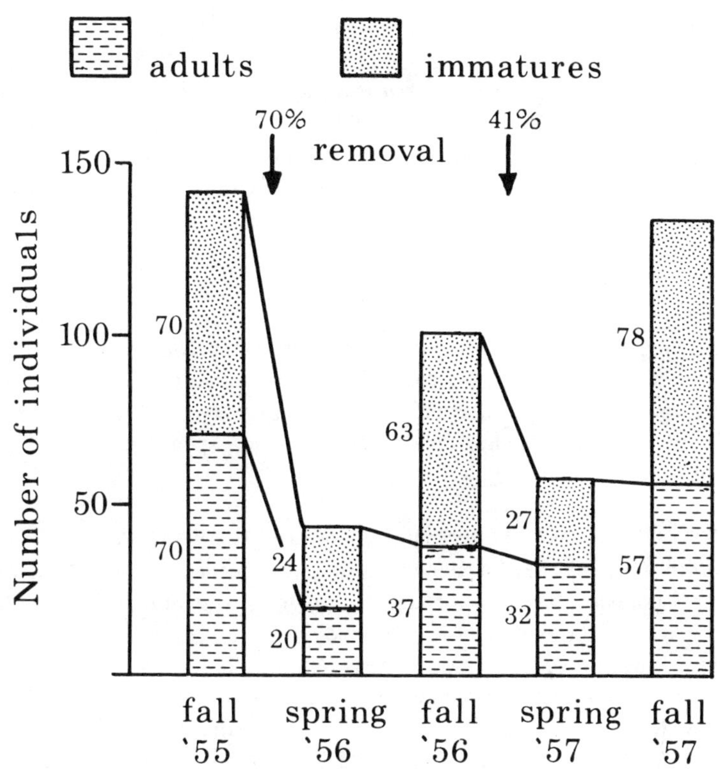

Figure 58. Recovery of a California Quail population in the Berkeley Hills after winter removal of 70% and 41% of the birds by trapping in two consecutive years. (After Raitt and Genelly, 1964.)

Others are densely grown to chaparral, which makes hunting virtually impossible. A good many ranches are posted and are hunted only lightly or not at all. Other areas close to roads and open to the public may be overhunted, with the coveys cropped heavily and the survivors badly scattered. But when the next breeding season comes around, pairs of birds from intact coveys scatter widely over the countryside and serve to repopulate local areas where the breeding stock may have been unduly reduced.

The best evidence that deployment of breeding stock does indeed occur is supplied by the study conducted in the Berkeley Hills by Raitt and Genelly (1964). After the quail population had been investigated for 5 years, and normal stocking and turnover rates were known, the authors removed by trapping 70 percent of the fall population in 1955 and 41 percent of the fall population in 1956. As shown in Figure 58, the populations completely recovered in each case to normal stocking levels. In fact, the population in the fall of 1957 was one of the highest recorded in 7 years. The age ratios of the recovered populations did not suggest that recovery had resulted from higher productivity of young per surviving adult. Rather, the banding data indicated quite clearly that there had been a marked influx of breeding pairs from adjoining range during the two springs following the heavy imposed mortality. The authors logically conclude that heavy mortality in one sector of a populated range may easily be absorbed by the population as a whole. It does not follow that mortality as high as 70 percent could safely be imposed on the total population without depressing the subsequent level of stocking.

MISCELLANEOUS LOSSES

Quail, like all other wild-living animals, are subject to accidents. Birds occasionally fly into wires or other obstructions and are maimed or killed. Traffic losses occur where quail fly across highways. Floods along river plains may drown some birds and force others into open terrain where they are subject to predation.

DENSITY-DEPENDENT VS. DENSITY-INDEPENDENT MORTALITY

Types of mortality may be categorized into two general classes. Predation, hunting, parasites, and diseases are all *density-dependent* mortality factors. That is to say, the likelihood of birds being lost from a population to any of these causes is highest in dense populations and becomes progressively less as populations decline. Small residue populations may persist in desert habitat through several dry seasons, when reproduction is precluded by virtue of lack of rainfall and proper nutrition.

Generally speaking, density-dependent sources of mortality are "compensatory," to use the term coined by Errington and Hamerstrom (1935, 1936) in their studies of Bobwhite ecology. Thus, if a substantial part of a fall population is removed by hunting, there will be fewer birds lost to predators. Conversely, if a population is not hunted, the surplus will still be dissipated by predation or other forms of mortality. A population cannot be stockpiled or accumulated beyond what the habitat will normally support. Thus, there are no more quail in a park or a refuge than in a conservatively hunted area, if the habitats are equal.

On the other hand, losses to poisoning by rodenticides or insecticides, to catastrophic weather conditions such as deep snow, floods, or drought, or to accidents are all *density-independent* and can induce mortality irrespective of whether the population is high or low. These sources of death are additive to other losses, and are not compensatory.

Photo by Paul J. Fair.

PART III
QUAIL MANAGEMENT

11
SUPPLYING COVER NEEDS

THE STRATEGY OF MANAGING QUAIL RANGE

A. Leopold (1933:vii) depicted the management concept as follows: "The central thesis of game management is this: game can be restored by the *creative use* of the same tools which have heretofore destroyed it—axe, plow, cow, fire, and gun. A favorable alignment of these forces sometimes came about in pioneer days by accident. . . . Management is their purposeful and continuing alignment."

This chapter and those that follow will suggest ways in which quail populations can be increased by the creative use of land management practices.

We must start with the realistic view that the management of lands and landscape within the range of the California Quail will be dictated primarily by economic considerations. Farming, grazing, logging, and other profitable extractive enterprises will continue to dominate the plans and programs of land utilization. If quail are to be encouraged as a by-product of land industry, management plans must coordinate logically with the primary uses. I see no basic conflict in this regard. The techniques of improving quail range are not expensive in terms of dollars or land. Rather, what is required is determination on the part of the *land operator* and continuous attention to the task. No one else can manage quail. It is an idle dream to think that a regulatory agency such as the Fish and Game Commission can effectively achieve the level of husbandry required to produce quail. These suggestions, therefore, are directed primarily to the rancher and to those governmental agencies that actually manage land—the federal

Forest Service, the Bureau of Land Management, and the military services in particular, and to those state and local agencies that own and administer land.

THE NEED FOR BRUSH COVERTS

Brushy cover is perhaps the most characteristic attribute of usable range for California Quail. As noted in Chapter 4, the daily routine of a covey of quail is designed to conform to the availability of safe points of retreat. The morning and evening feeding sites are adjacent to escape coverts. As the birds move from one part of their range to another, there must be safe enclaves along the way. During the day, coveys will usually loaf in the densest and most secure shelters available. At night, the birds need dense cover well above the ground for roosting. And lastly, during the season of reproduction, the pairs seek a very special type of ground cover for concealment of nests. A program of quail management must attempt to supply all the types of cover needed by quail, well distributed over the management unit.

Fortunately, the gross extent or area of cover required is relatively modest. Far more important is the quality and density of cover, and its spatial distribution.Excellent quail populations can be sustained in relatively open country by developing coverts and travel lanes that occupy only one or two percent of the total landscape.

Of the millions of acres of brush and scrub situated within the range of the California Quail, relatively little is in optimum condition for quail occupancy. Virtually all brushlands are grazed by domestic livestock, and the usual situation is that the ground between individual shrubs is bare and devoid of low cover suitable for hiding a quail. Thus, in the foothills there are extensive shrublands of buckbrush (*Ceanothus cuneatus*) or coyote bush *(Baccharis pilularis)* that from a distance would appear to be ideal for quail. Upon close examination, however, the ground level is often found to be sorely deficient in hiding spots where a bird could escape the determined attack of a Cooper Hawk. The same can be said for the endless fields of big sage (*Artemisia tridentata*) in the Great Basin, or the rolling hills fringing the San Joaquin Valley covered with a mixture of xerophytic shrubs such as chamise (*Adenostoma fasciculatum*), rabbit brush (*Chrysothamnus* sp.), monkey flower (*Mimulus* sp.), California sage (*Artemisia californica*), or quail brush (*Atriplex* sp). To increase the usefulness of brush stands for quail, there must be cover at the ground level as well as overhead.

On arid ranges, often the most effective way to achieve this end is to excude livestock from portions of the brush. Elimination or close regulation of grazing permits the grass and weeds to form cover on the ground and likewise encourages the shrubs to spread horizontally in umbrella form, thereby adding effective shelter from above. Gullies and odd corners of brushy pasture thus protected from overgrazing can be converted from poor to excellent quail cover by construction of a bit of fence.

Figure 59. Fenced gully, well grown to quail cover and at the same time protected from erosion. I. McMillan ranch, Shandon.

Particularly crucial are areas of dense escape cover situated near water. In pasturelands, the riparian zone along a stream or near a spring is likely to be badly overgrazed by cattle that spend much time near the drinking source. The water is also needed by quail, but if cover is deficient the birds hesitate to venture across the open zone to drink. Fenced plots near a water source are quickly grown to dense vegetation which fills the needs of the quail with very little sacrifice of grazing area for livestock (see Fig. 73).

BRUSH CLEARING

There is a notable trend toward large-scale brush removal as a component of the clean farming syndrome. "Progressive" farmers view brushland as wasteland, and the idea is widely advertised that brush clearing is a proper step toward improving livestock pastures. Such clearing is highly deleterious to the welfare of quail. McMillan has appraised a typical and most graphic demonstration of brush removal near his ranch at Shandon. The site of this particular range improvement project was a big patch of chamise chaparral, some 1,200 acres in extent, that was somewhat isolated from other nearby areas of shrub cover. This unique piece of shrubland had long been well-known as the central habitat of one of the most stable and abundant quail populations in the entire region. Artificial watering sites for quail (guzzlers) had been constructed within this immediate territory, and there were other sources of water for wildlife. In the summer of 1973, with the chaparral and small trees first crushed by a mechanical technique, this big patch of quail cover was burned clean. In a few hours, more quail country was destroyed in this typical brush-conversion project than had been established by planting in the general area in 30 years of extensive quail restoration work.

Lawrence (1966) accurately documented the impact of a brush removal program on a 1200-acre property near Glennville, in Kern County. He censused the bird and mammal populations on the area prior to burning and for a period of 3 years following the fire. As would be expected, both birds and mammals typical of the chaparral decreased substantially in abundance, although grassland species increased. In the case of California Quail, the pre-burn population was reduced by half after the fire.

Moreover, one may question the validity of the assumption that most chaparral can successfully and profitably be converted from brush to grass to the benefit of the livestock industry. The chaparral still existing in California occurs largely on poor soils incapable of sustaining good stands of grass under the pressure of grazing. The stability of the soil and the watershed cover may be best protected by maintaining the well-adapted chaparral cover, particularly on steep slopes.

Another land practice now in vogue is the elimination of fences, and hence of brushy fence-rows, along field borders. This operation gains a few feet of cropland at the expense of many kinds of wildlife, including quail. I am increasingly distressed at the progressive "cleaning up" of field borders as a concomitant of modern, slick, mechanical farming. In proposing a policy of quail management, we must assume that the land operator is willing to dedicate a very limited area of the ranch to supplying minimum cover needs for quail occupancy. Fence-rows are among the most valuable of quail coverts because they serve as safe travel lanes for the birds to reach isolated foraging areas.

Figure 60. Extensive brush removal to improve livestock pasturage severely limits the usefulness of an area to quail. Only the edges of this large clearing are habitable quail range. U.S. Forest Service photo.

By the same token, the unnecessary elimination of brush growth along roadsides is a continuing threat to quail welfare. Most counties have road maintenance crews whose duties include tidying up the roadside by cutting, burning, or spraying the brush. The justification of this expensive and recurrent activity is by no means clear. The adverse effect on wildlife of many kinds, however, *is* clear. Perhaps county road crews might better be assigned the task of *planting* appropriate and decorative shrubs to improve the attractiveness of roadsides, just as shrubs like oleander and pyracantha are being cultivated along the shoulders of major trunk highways for roadside beautification. The benefits to birdlife, including quail, would be substantial.

A first step in cover management for California Quail certainly should be the maintenance of existing coverts on areas not needed or not suited for more important uses.

BRUSH MANAGEMENT

There are some situations where extensive areas of dense brush may profitably be broken up by clearing lanes or strips, primarily for improving livestock forage and for fire control, with substantial benefits to quail habitat. The central portions of a brush field may supply excellent cover but are of limited use to quail in respect to other living needs. Small clearings permit the growth of quail food plants and extend the border zone or ''edge'' between cover and foraging ground. Such modifications of existing brushlands, however, are of a character very different from the wholesale brush removal programs referred to above.

One of the most effective management programs for California Quail has been developed by Ray Conway in the oak woodlands of western Yuba County. Much of the foothill region east of Marysville had been repeatedly burned, giving rise to thickets of live oak, black oak, and various shrub species that were unproductive for either quail or livestock because of inadequate ground forage. Conway developed a system of opening up the cover with a bulldozer, pushing the tall saplings into brush piles, and seeding the openings so created with subterranean clover and rose clover, after fertilizing the soil with super-phosphate. The brush piles serve initially as cover, and the clovers plus other weedy forbs supply food for both quail and winter/spring livestock grazing. When the brush and tree cover resprout, supplying more natural quail cover, the large brush piles are burned and the ash beds are seeded to clover. Thereafter, the distribution and density of brush is regulated by periodic bulldozing, creating an optimum dispersion of low brushy cover and feeding areas. Small brush piles created by the bulldozer serve as supplementary cover. Some dense live oaks are left for roosting cover, and an interspersion of healthy black oaks is retained for acorn production and general attractiveness of the landscape.

The result is an open oak woodland with scattered large trees, small brush piles, live oak sprouts, and a healthy growth of grass and forbs in the openings. Cattle are grazed from November to May, but are removed before the summer drought to permit the clovers and forbs to set seed. About 500 to 600 cows plus calves are grazed on the 4400 acres under management. The quail population has been increased from a negligible level to an average fall density of 2 birds per acre. Approximately two-thirds of the income now derived from the property comes from livestock, the other third from sale of quail shooting rights. The area has become an important winter range for migratory deer and a year-round range for some wild turkeys and numerous cottontail rabbits. Coyotes and other predators are tolerated, to assist in controlling ground squirrels and other rodents.

The obvious object lesson to be derived from Conway's management plan is that brush management and cattle grazing, used with judgement and skill, can produce range that is excellent for both cattle and wildlife. The widespread fetish that removal of trees and brush must be complete to improve livestock forage is an unfortunate myth.

PLANTING ESCAPE COVER

In cover-deficient areas, it is sometimes necessary to introduce cover plants artificially. Thus, in the arid hills near Shandon, McMillan has developed

Figure 61. Managed brushlands in the oak belt east of Marysville, Yuba County. Note the brush piles and dense oak sprouts. Ray Conway harvests over a quail per acre on some of these lands.

highly successful quail coverts by planting hedges and small blocks of atriplex, variously called saltbush or quail brush. The endemic species, *Atriplex polycarpa,* is a hardy, drought-resistant shrub that develops dense cover within three or four years after planting. Two other species, *A. canescens* and *A. lentiformis,* are equally well-adapted to the San Joaquin foothills, though not native there. Plantings of quail brush must be protected from excessive grazing, however, since all species are highly palatable to cattle and sheep during the dry season, when other forage is scarce. When protected from livestock, quail brush develops dense lateral foliage that droops down to ground level, and even an isolated bush can offer secure protection from predators or from weather. A strip of atriplex along a field border, protected by parallel fences, can harbor a covey of quail in an area otherwise devoid of cover and useless as quail range.

McMillan (1960) has recorded the methods he has used successfully in establishing atriplex hedges for quail in the Shandon area. Seeds can be gathered from an existing bush by shaking the heavily laden branches over a canvas. The seed bed is prepared by discing or otherwise cultivating in late winter (January) to kill the sprouting winter annuals. Seeds are scattered on the surface of the bare soil, and the strip is then lightly dragged to incorporate the seeds into the surface layer. A thick stand of seedlings will result from application of 100 pounds of seed per acre. There must be adequate moisture to assure germination. A program of propagation should include provision for replanting, in case of initial failure in a dry year. Once sprouted, the seedlings compete vigorously with annual vegetation, and no further cultivation or care is required except protection from livestock grazing and fire. A functional, self-perpetuating hedge will develop after the third year of growth. By the fourth year, the canopy will be adequate to serve as quail cover (see Fig. 62).

In moist situations, common blackberry (*Rubus laciniatus*) and Himalayan blackberry (*Rubus discolor*) are among the most effective plants to harbor and protect quail. They are hardy and easy to establish by planting root suckers. On the Penobscot Ranch near Georgetown in Eldorado County, virtually every covey of California Quail makes its headquarters in a clump of tangled Himalayan blackberry vines, usually situated along a stream or beside a hillside seep. Some clumps crucial to individual coveys are no more than 20 to 30 feet in diameter and perhaps 5 to 6 feet high. Parts of the Penobscot Ranch where no blackberries exist support very few quail, although in all other respects they appear to be as favorable as the well-populated sites. On the Alan Starr ranch near Mission San Jose, Alameda County, a blackberry hedge bordering the entrance road supported a large covey of birds for a number of years. When a new foreman, in a campaign of tidiness, hacked out the hedge, the covey disappeared. But the roots were hardy, and within two years the berry canes began to re-establish the hedge, and quail re-occupied the site. One advantage of the blackberry is that it

Figure 62. Atriplex hedges on the McMillan ranch, Shandon. Following these ribbons of safe cover, quail forage widely over the area. Junipers along the arroyo are used for roosting.

Figure 63. Well-distributed blackberry clumps support a high quail population on Penobscot Ranch, Georgetown, El Dorado County.

defies grazing and does not have to be protected by fence. One big disadvantage is that it continually spreads and if not periodically killed back will ultimately occupy a whole meadow. Dutson (1973) documents the spread of the Himalayan blackberry in moist pasturelands of northern California. Control, when necessary, is best achieved by the local application of herbicides. Large berry patches, inpenetrable to man or dog, can be broken up to facilitate quail hunting by pushing down swaths with a bulldozer.

Another outstanding cover plant is the tamarisk *(Tamarix gallica)*, introduced originally from Europe. Along river courses or desert washes the hardy tamarisk or "salt cedar" grows densely and profusely, and if ungrazed drops its canopy to the ground like a skirt. Even a single luxurious plant can constitute a covey headquarters, supplying escape cover, loafing cover, and a roost site.

Conifer plantations, when young and not too tall, serve as excellent quail cover. If planted in narrow strips, the low lateral limbs stay alive for many years maintaining shelter at ground level where it is needed.

I have named just a few examples of the many plants that might be established for quail use. In watered yards or other moist sites, a myriad of native and/or cultivated plants may serve the needs of quail while at the same time adding attractiveness to the landscape. Emlen and Glading (1945) list 46 shrubs that can be utilized in a quail management program.

ROOSTING COVER

California Quail will roost in any dense plant with evergreen foliage that offers concealment from owls and enough elevation to be above the reach of ground predators. Live oak (*Quercus agrifolia*), bay tree (*Umbellularia californica*), common juniper (*Juniperus californica*), and various conifers (pines, fir, redwood, etc.) are among the roosts most commonly chosen by California Quail. Where these are present naturally, roosting cover may be completely adequate. On the other hand, there are many situations in arid rangelands where roosting sites are few or absent. Junipers and live oaks are easily planted to fulfill this need, but both require protection from grazing to achieve maximum growth. Some of the shrubs and vines mentioned previously may grow tall enough for roosting, particularly *Atriplex lentiformis* and Himalayan blackberry. In Baja California various kinds of cacti, most especially nopal, are used by quail as safe roosts.

Growing roost plants where none exist takes time, and it may be desirable to supply artificial roosting cover in the interim. For this purpose the "elevated brush pile" developed by McMillan serves admirably. Four 8-ft. steel posts are driven in the ground, spaced 8 ft. apart, and a platform is constructed using additional steel posts lashed to the tops of the four corner posts. The platform then serves as a base for limbs and brush that can be piled to construct a roosting shelter elevated 6 ft. or so above the ground. When such a roost is situated in an area otherwise attractive to quail, the

Figure 64. Artificial quail roosts, Ian McMillan Ranch, Shandon.

birds will quickly discover and use it (see Macgregor, 1950). McMillan has devised other effective plans for roost construction. Two parallel cables stretched across a gully can serve as a base for an elevated brush pile, or the shelter can be built in the crown of a living tree. Several hundred birds will crowd into one of these artificial roosts if it is the only roosting site available.

BRUSH PILES

As already recounted, Conway uses brush piles loosely assembled by the bulldozer as interim cover during the process of developing natural stands of low brushy cover. Emlen and Glading (1945) depict another design for brush pile construction. Logs or heavy poles are laid as a base, and the pile is thus elevated from the ground, giving space beneath for the quail to move around. An old farm gate or a wooden frame can be used in similar fashion to support the brush. Such an elevated brushpile will be slow to settle and rot because the limbs are not in contact with the ground. Orchard cuttings, tree trimmings, or tops from a logging operation may be used to construct the pile. The brush piles serve the quail as impregnable fortresses, offering escape and loafing cover and even roost sites in areas devoid of proper roosting trees.

In a hunted quail population, brush piles are advantageous in stopping birds from running away from the hunter. After a covey is flushed, the scattered birds will tend to take shelter in the brush piles rather than to alight and run, as is often the case when ground cover is sparse.

On his ranch near Shandon, McMillan uses small brush piles to encourage quail to forage several hundred yards into open fields that would otherwise be unusable. The large "home covey" of 800 to 1400 birds exhausts

Figure 65. An oak tree, torn to the ground by heavy snow, creating excellent quail cover. A freak storm on January 4–5, 1974, felled thousands of live oaks in an area 70 by 20 miles in the Monterey Mountains, substantially raising the carrying capacity for quail. J. Davis photo.

the food supply close to atriplex cover early in the winter, and the birds are forced to extend their morning foraging farther and farther into open country. Movable brush piles facilitate these forays into coverless terrain.

In January of 1973, a heavy, wet snow fell on the coast ranges from Monterey to San Luis Obispo, and enormous numbers of live oaks were broken off and fell to the ground by the weight of the snow on the foliage. These "natural brush piles" created highly favorable cover for quail in areas where cover had been inadequate before. At the date of writing (1975), the leafy tops of these fallen oaks are still serving as cover bastions for both California and Mountain Quail.

The primary disadvantages of depending on brush piles for quail cover are: (1) they are temporary and eventually disintegrate through rotting; and (2) they are highly vulnerable to destruction by wildfire.

OTHER TYPES OF COVER

A technique of creating "living brush piles," which was devised originally on the King Ranch in Texas for Bobwhite management, has been effectively used for California Quail. A bushy live oak, preferably under 4 inches in diameter, can be partially severed so that it falls to the ground while retaining a strip of living bark connecting the stump to the top. In a prone position the tree will continue to live, sending out a thicket of sprouts that form excellent cover at ground level. A number of such enclaves of cover were developed with great success on the Penobscot Ranch in El Dorado County, where the author hunted quail for some years. The technique can be used with a number of trees (willow, mesquite, various oaks, etc.) and tall shrubs (coffee berry, toyon) that are not so brittle as to break off cleanly when the top is felled.

The Russian thistle or tumbleweed tends to form piles or windrows along fences or in gullies. It is common practice to burn these accumulations in accordance with clean farming practice, but much usable cover for quail, rabbits, and small birds is thereby wasted. By the simple device of leaving these piles unburned, the cover value can be preserved and soil erosion retarded.

In some areas of quail range, rock piles and outcrops constitute important cover for quail and also for chukar partridges. A rough pile of rocks discourages excessive grazing and protects vegetation in the interstices from consumption by livestock. On the San Joaquin Experimental Range in Madera County, rock outcrops constitute a crucial part of the quail cover. Similarly, in parts of eastern Oregon, Washington, Nevada, and Baja California, rocky areas are regularly used by California Quail.

In a well-developed quail range, units of cover should be no farther apart than a quail can fly comfortably, which is 200 to 300 yards. Optimum spacing would be even closer, but in farming country this is not always possible. Long narrow strips of cover, such as hedges or fencerows, are more useful than isolated blocks since the birds can use these as safe avenues for foot travel.

NESTING COVER

The essential component of a suitable nesting site is herbaceous ground cover (grass or weeds) sufficiently tall and dense to conceal the nest. In pasture lands, such clumps of ground vegetation are usually situated at the base of a bush or in a fallen tree or limb that precludes slick grazing by livestock. The best areas for quail nesting are portions of more or less open grassland that are grazed lightly, or in arid zones not at all. Some nesting sites are supplied in fence-rows, road ditches, and rims of gullies that are protected from livestock. These narrow areas, however, tend to serve as travel lanes for ground predators that destroy many quail nests. The safest nest sites are in wider units of ground cover that tend to randomize the passage of predators. Seasonal grazing is not deleterious since new cover for use the following year will be restored again in the subsequent spring's growth. Regulation of grazing, therefore, is the key to supplying nest cover, as is the case in maintaining other types of cover and food resources as well.

THE SIGNIFICANCE OF COVER IN QUAIL MANAGEMENT

As stated at the outset of this chapter, the California Quail is basically a bird of the brushlands, and dense woody cover is essential to sustain high populations. I emphasize again the importance of *density* or cover quality, in contradistinction to mere extent of brushland. Unless cover is thick at ground level and impervious to penetration by avian predators such as the Cooper Hawk, it will not adequately serve the needs of quail.

Programs of land management that call for wholesale removal of brushy cover are in essence programs of quail elimination. Protecting crucial units of quail cover from complete elimination or damage by excessive clearing, burning, spraying, or overgrazing is the first essential step in developing a plan of quail management. In areas supplied with a plethora of cover, such as wide expanses of chaparral or continuous forest, the creation of interspersed openings will benefit the quail. Where natural cover does not already exist it may have to be planted.

12

MAKING FOOD AVAILABLE

FOOD AS A LIMITING FACTOR

Within the presently occupied range of the California Quail, inadequacy of a year-round food supply is another important factor limiting population levels. Just as lack of proper cover depresses quail numbers in some areas, the lack of reliable food supply becomes the deterrent in others. Extensive areas of foothill rangeland reasonably supplied with vegetative cover and with scattered but reliable sources of water are supporting low quail densities because of periodic food shortage. It must be remembered that carrying capacity of quail range is dependent on *all* environmental requirements being met throughout every season of the year. There are many areas where food may be more than adequate for part of the year but deficient in certain seasons. The average quail population will be keyed to those periods of minimum food supply, following Liebig's "Law of the Minimum."

Emlen and Glading (1945:29) point out that "Quail food shortages are of two types: (1) acute shortages of brief duration resulting from snowstorms or floods, or from the sudden destruction of major feeding areas by cultivation or other drastic treatment; and (2) chronic shortages resulting from more or less continuous deficiencies in the vegetation because of unfavorable soil, climate, or land management."

Acute shortages are rare in California and Baja California because of the low incidence of snow and ice. However, in the northern and eastern portions of the species range, periodic severe winter storms may render all food unavailable, leading sometimes to starvation and substantial losses of quail. Floods are usually too localized to result in actual starvation, although they may displace coveys situated along riparian bottomlands.

Much more common in California, and elsewhere in the species range, is chronic food shortage stemming from one or the other of the following causes:

1. *Overgrazing of rangelands.*—Within the arid and semi-arid zones occupied by California Quail, overgrazing by livestock is the most common and significant factor limiting quail food supplies, and hence quail populations. The foothill pasturelands which constitute the main stronghold of the quail are quite generally grazed to a degree that lowers the carrying capacity through adverse effects on both food supplies and cover.
2. *Clean cultivation of croplands.*—As noted in Chapter 4, the rich tillable valley lands are cultivated so intensively as to be essentially uninhabitable for quail. Food in the form of crop residues may be produced in local abundance, but these supplies are seasonal rather than sustained, and in any event lack of cover makes this potential resource of little value to quail.
3. *Dense growth of woody vegetation.*—Excessively dense vegetation precludes the growth of forbs that constitute the mainstay of the quail diet. In mountainous areas, especially in the humid coastal zone, a closed canopy of forest or chaparral suppresses the growth of forbs at ground level. Whereas cover may be more than adequate, food supply often is not.
4. *Infertility of soil.*—Irrespective of patterns of land use, wildland soils are not uniformly productive of forb growth. I know from personal hunting experience that coveys have certain areas where they feed regularly, whereas other equally available sites are rarely visited even though the floristic composition appears to be similar. There is strong circumstantial evidence that preferred feeding sites are more fertile, and that quail are attracted to such sites either because of higher seed production or through recognition of qualitative differences in the food produced.

The first three causes of chronic food shortage are subject to some degree of modification or correction; soil deficiency is not.

FEEDING HABITS OF CALIFORNIA QUAIL

Upon descent from the roost at daylight, a covey of quail will move purposefully to a suitable feeding area and begin foraging. Within an hour the crops are half-filled or more, and feeding intensity decreases. Desultory feeding may continue through the morning, and by mid-day the covey is usually found loafing quietly in dense cover. Active feeding is resumed in late afternoon, and there is a burst of intense foraging activity as evening approaches. The birds ordinarily go to roost with crops bulging.

Stormy weather interferes seriously with the normal feeding regime. In steady rain or snowfall, quail are reluctant to leave shelter and may spend most or all of the day huddled in a brush pile or dense thicket. Even in the worst weather, however, the birds venture forth to forage before dark, apparently being loath to go to roost with empty crops.

The presence of a natural enemy, such as a Cooper Hawk, will keep quail confined to cover and reduce the effectiveness of feeding. Just before dark, however, the birds are driven to attempt to feed, even at great risk of being captured. Sumner (1935:185) states: "Possibly it is for this very reason that the Cooper Hawk is especially active in the pursuit of quail at dusk."

During the breeding season, the feeding program is substantially altered. The cock bird is preoccupied with driving away intruding males and standing watch over his hen and her nest, with the result that he does little foraging during the day. The female who is incubating eggs leaves the nest for a short period of feeding in the early morning and intermittently thereafter, always ending the day with her crop at least partially filled. After the young hatch, and for a period until the chicks can fly to roost, the parents are busy tending the brood, and they subsist on short rations. For this reason, both the cock and the hen lose weight during the stressful period of reproduction (as shown in Figure 31, Chap. 6).

Generally speaking, quail forage only in areas adjacent to escape cover. Sumner (1935:279) states: "Although exceptions are sometimes observed, there is a fairly definite cruising radius from cover of about fifty feet, beyond which the birds usually will not venture even when pressed by hunger, so great is their aversion to open places." Emlen and Glading (1945) repeat this estimate of the extent of usable foraging ground in relation to cover. At Shandon, McMillan has observed that in late summer and early fall the birds will venture out several hundred yards to forage, but with the autumn arrival of the first migrant Cooper Hawks the feeding radius immediately shrinks to the close environs of escape cover. The obvious implication is that in open regions, extending the distribution of cover is the most effective way of augmenting the availability of feeding grounds, and hence of food. In chaparral or other areas oversupplied with woody cover, the opposite is the case—namely, openings where food may be produced are in short supply. The objective of management is to maximize the interspersion of cover and food-producing open areas.

SEASONAL CHANGES OF DIET

The California Quail is an opportunistic feeder, and dietary intake will vary substantially from one locality to another, and within one area from season to season and from year to year. Quail food habits have been much studied, and there is an extensive literature based on analysis of crop contents. Some of the more important contributions are papers by Anthony (1970b), Crispens et al. (1960), Glading et al. (1940), Shields and Duncan (1966), Sumner (1935), and Appendix B of this volume.

The main components of the California Quail diet are seeds and green leafage. Insects are taken only casually in spring and early summer—more in wet years than in dry. Figure 66 denotes in diagrammatic form the average seasonal consumption of seeds, greens, and insects, based on data presented in the tables in Appendix B. The foods designated collectively as "seeds" include fruits, berries, acorns, and crop residues, in addition to the seeds of forbs, but the latter constitute the bulk of the food volume at all times of year.

In the coastal ranges of California, Mediterranean bur clover (*Medicago hispida*) is perhaps the most important single food plant. The seeds are eaten virtually throughout the year, either encased in burs or as individual seeds when the burs soften and degenerate. Bur clover leafage is taken also in large amounts during the growing season. Sumner (1935:178) emphasized the significance of this important forage plant, which constituted 14.5% of the annual diet of quail in Santa Cruz County. In analysis of 102 crops, Sumner found the average number of bur clover seeds to be 185, the maximum number 1224.

In the more arid interior foothills, other legumes take the place of bur clover. Various clover species (genus *Trifolium*), lotus (*Lotus*), and lupines (*Lupinus*) are dominant dietary items at Shandon. Fully as important as the legumes are species of filaree (*Erodium*) introduced from the Mediterranean region at the time of Spanish settlement. Glading et al. (1940:134) studied quail food habits in Madera County, and they state: "Plants of this genus, chiefly *Erodium botrys,* constituted the largest single food item in the annual diet, making up 20.4 per cent of the total." Filarees are common components of the annual vegetation of the California foothills, and the California Quail profits enormously from the introduction of these exotics.

Other annual forbs that contribute significantly to quail sustenance are fiddleneck or buckthorn weed (*Amsinckia*), turkey mullein (*Eremocarpus*),

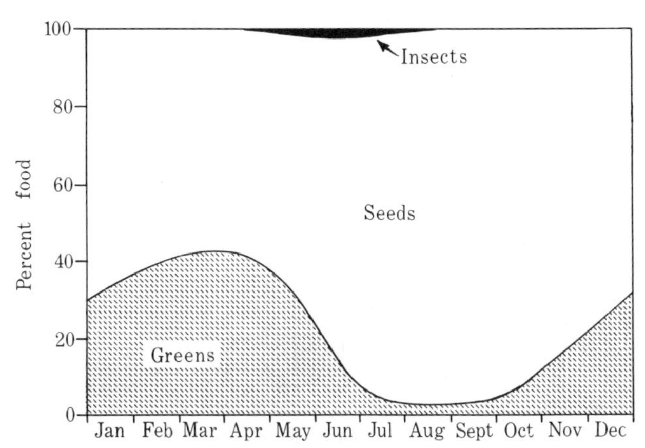

Figure 66. Simplified chart of the yearly diet of California Quail, based on data presented in Appendix B.

geranium (*Geranium*), vetch (*Vicia*), and various thistles (*Centaurea*, etc.).

By and large, the annual grasses that predominate over much of the California rangeland are not important quail foods, and in fact the grasses compete with and displace the broad-leaved annuals enumerated above. Wild oats (*Avena*) and soft chess (*Bromus*) are two of the worst in this regard. They contribute modestly to the quail diet, but not in proportion to the ground they occupy.

Of woody perennial plants, the oaks (*Quercus*) are much the most important contributors to the quail diet. Acorns are a rich source of nutriment at times and places when a good mast crop is produced. Some other woody plants whose fruits are taken on occasion are poison oak (*Rhus*), manzanita (*Arctostaphylos*), elderberry (*Sambucus*), and acacia (*Acacia*). The beans of black locust (*Robinia*) are eagerly sought by quail during periods of snow and cold.

GREEN FOODS

Many of the plants whose seeds sustain quail also contribute to the green leafage that is consumed seasonally. Important among the producers of edible greens are the legumes, including bur clover, other clovers, lotus, and lupine. Other annual forbs supplying green forage are filarees, chickweeds, mustards, buckthorn weed, and miner's lettuce.

The consumption of greens by California Quail begins in the fall, shortly after the first rains. Browsing peaks in winter and spring, and virtually ceases in summer when the annual plants die and the current crop of new seeds becomes available. Figure 67, derived from the tables in

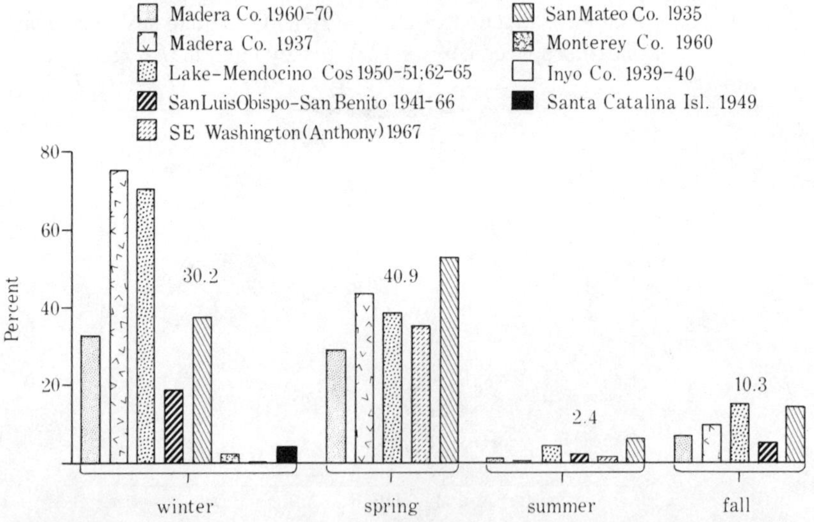

Figure 67. Seasonal occurrence of green leafage in the diets of California Quail sampled in various parts of the species range. (For details see Appendix B.)

Appendix B, shows the general seasonal trends in the use of green foods. It also shows how variable this food habit can be. In winter, for example, greens may constitute as much as 75% of the quail diet, or as little as 0 to 3%.

The significance of variable intake of greens is not completely clear. Sumner (1935:181) expresses the view that quail take greens only when the supply of seeds is inadequate. He states: "So long as the seed supply holds out it seems to be preferred to green leaf materials, and doubtless it is much more nutritious per unit of volume. The writer has seen penned birds gradually starve to death on a diet of green feed alone, while the birds showed an extraordinary eagerness for seeds whenever these were supplied. However, as the seed supply approaches exhaustion in January and February, the birds are forced more and more to depend upon green food. . . ."

On the other hand, it seems significant that quail start to browse immediately when greens are available, even in years of seed abundance. Furthermore, as noted in Chapter 9, Erwin found a high intake of greens in the winter and spring of 1973 when reproductive success among the quail at Shandon was exceptionally good, and relatively low consumption in 1972 when reproduction was minimal. It seems more likely that the *palatability* of green leaves varies from year to year, in accordance with the vigor of growth and the chemical composition of the leaves. When forb growth was stunted in 1972, the leaves were shown to contain various isoflavones which seemingly curtailed reproductive activity and which also may in some way be distasteful to quail. Robust forb growth, as occurred in 1973, was eaten avidly by the birds (up to 73% volume of leaves in the birds' crops), and qualitative analysis showed a very low incidence of isoflavones.

Isoflavones are only one category of chemical compounds that vary in occurence in green leaves. There may be many other compounds that affect palatability, nutritive value, and reproductive stimulation or inhibition in animals that consume them. I still suspect the existence of some substance in vigorously growing plants that may stimulate reproduction, just as isoflavones occurring in stunted growth inhibit reproduction. But such a stimulant has not yet been identified.

In the coastal and more northerly parts of the California Quail range, the clovers and other choice greens seem to be available and palatable every year, so quail breed annually, although with varying success. The "boom or bust" regime is most apparent in the arid southern portion of the California Quail range—the San Joaquin Valley and adjoining foothills, the western fringe of the Colorado and Mohave Deserts, and throughout Baja California.

Among the tables in Appendix B will be found occasional entries entitled "grass leafage." Quail eat green grass in large quantity only when other foods are scarce or absent. Sprouting green grass is available every spring in virtually all parts of the California Quail range, and it is eaten

Figure 68. Wild legumes such as *Lupinus bicolor* supply quail with green leafage in winter and spring, and seeds year-round. Jepson Herbarium photo.

regularly in small quantities. Seeds or sprouting forbs seem to be preferred, however, when they are readily available.

In eastern Washington, Anthony (1970b) noted some differences in the amount and kinds of green foods eaten in spring by female and male California Quail. Hens were particularly avid for the green flower heads of jagged chickweed (*Holosteum unbellatum*). They ate virtually no wheat (seeds), whereas males ate a good deal of wheat and a wide spectrum of greens, including chickweed. Perhaps the physiologic demands of the two sexes are differently expressed in terms of dietary preferences during the period leading up to nesting.

AGRICULTURAL CROP RESIDUES

In summer and fall, California Quail make good use of agricultural crop residues left on the ground after harvest. Wheat, barley, and corn are the principal crops used. In the Central Valley, rice, kafir, and safflower are available locally, where remnants of riparian cover sustain a few quail in heavily developed agricultural communities.

However, the value of crop residues in supporting quail on a sustained basis is limited by the fact that assured availability is restricted to summer and fall, when weed seeds are generally abundant. During the critical period of winter and early spring, when quail food may be in short supply, crop residues disappear as stubbles are plowed under. In the San Joaquin foothills, wheat and barley stubbles customarily are fallowed in autumn to catch and hold winter rains in the soil. Even in unfallowed stubbles, the seeds soften and sprout with the first heavy rains. Similarly, in the Central Valley, stubbles of corn, rice, and safflower are turned under in dry autumns to prepare the seed bed for sowing the following spring. In short, this source of food is too unreliable to contribute importantly to the basic carrying capacity of a quail range. Observation at Shandon suggests that the sustained breeding stock of quail, carried through the winter on a year-to-year basis, is little affected by agricultural residues but depends much more directly on the supply of forb seeds and greens.

GRAZING IN RELATION TO FOOD AVAILABILITY

As stated previously in this chapter, grazing by domestic livestock is the primary factor limiting the production of quail foods in the arid foothill ranges of California. In spring, when the weeds and forbs are sprouting and flowering, they are generally highly palatable to domestic livestock. Cattle in particular will closely graze succulent forbs, especially the important legumes, reducing or eliminating the production of a crop of seeds. Even plants that set seed may be consumed later in the summer before the seeds fall to the ground. The result on heavily grazed ranges is that the seed supply available for winter use of the quail is only a fraction of what it might be with light or moderate grazing.

The degree of impact of grazing on forb growth is generally an inverse function of rainfall during the previous winter and spring. In a dry year, forb removal may be virtually complete, whereas on the same range in a year of good rainfall the forbs grow faster than the livestock can eat them, resulting in a good seed drop despite grazing. The dry years, of course, are the critical years for quail survival.

Close grazing has yet another adverse effect on seed availability. Where weeds and grass are permitted to grow and cast a layer of duff or decaying vegetation on the ground surface, weed seeds are held in the duff and may be obtained by the quail through scratching. When, however, all current annual growth is consumed by grazing, there is no duff layer, and such seeds as may be produced become imbedded in the bare soil, or are consumed by rodents and small birds before the quail can use them. Lehmann (1953:240) noted the storage function of duff in holding seeds where Bobwhite Quail could find them. He states: "Long after the heavy combined pressure of seed eaters had largely removed the seed from the naked land, quail in tall grass were still finding seed . . . protected from quick dissipation by . . . plant debris." An optimum quail range should retain enough surplus vegetation to form a duff layer, at least under the sheltering canopy of shrubs if not in openings.

Within a fenced paddock, grazing intensity is rarely uniform. It is normally heaviest near water and on flat areas or gentle slopes, lightest in corners far removed from water and on steep terrain. Improving stock water distribution on rangeland serves often as a mechanism for spreading the scourge of overgrazing. Ironically, various so-called conservation programs of federal agricultural agencies subsidize the development of new livestock watering points on rangelands, justifying the cost in part as "wildlife habitat improvement." The effect on wildlife range may be just the opposite of improvement.

Conversely, in areas of moderate to high rainfall, grazing may be the most practical technique for thinning dense grass stands, permitting the growth of seed-producing forbs. In the Coast Ranges and Sacramento Valley foothills, some ungrazed rangelands are poor in quail food and poor in quail.

There are various ways in which grazing can be regulated with a view to quail range improvement: (1) fence portions of a property to preclude grazing entirely; (2) regulate stocking in terms of numbers, or seasonal use, or both, to protect key areas of quail range from overgrazing; (3) limit water availability for livestock to reduce grazing pressure in areas of dry range; and (4) utilize livestock to open up dense grass stands in areas of higher rainfall. In field practice, all of these techniques are practical, alone or in combination, depending on the local situation.

On both public and private ranges, the purposeful regulation of grazing is a key element in a management program for the California Quail.

OTHER CULTURAL PROCEDURES TO STIMULATE FORB GROWTH

Assuming that grazing is under control, there are various forms of land treatment that can be applied to augment supplies of forbs for quail food.

Solid stands of grass, if broken up with a disc harrow, usually will produce a heavy growth of forbs and weeds as initial stages in plant succession. Filaree, lupines, lotus, clovers, and buckthorn weed are among the volunteers that spring up on bare soil. McMillan uses this technique regularly to improve quail food, especially in strips adjacent to hedges of *Atriplex* or other coverts. At times, however, disturbed soil will perversely produce only more grass, although this happens rarely.

To a lesser extent, forbs can be encouraged to sprout by early spring burning of spots or narrow strips in heavy stands of annual grass such as wild oats. This procedure likewise is used successfully on the McMillan ranch in limited situations.

AUGMENTING FOOD IN CHAPARRAL AREAS

As noted in the last chapter, extensive unbroken areas of chaparral offer to quail a plethora of cover but a deficiency of food. This is not universally the case, for in some years of favorable rainfall forbs will grow in considerable quantity in areas where the canopy is thin or broken with small openings. Ordinarily, however, dense tangled stands of chaparral, brush, or scrub forest are characterized by food scarcity, and quail must forage in adjoining clearings or openings more amenable to forb growth.

Emlen and Glading (1945:24) propose that chaparral stands can be broken up by clearing 30-foot swaths with a bulldozer. They state: "Extensive stands of brush should be broken up into small patches by clearing lanes through them. Once the brush has been removed, annual plants useful as quail foods usually appear quickly, especially where some soil has been turned up." On pages 16 and 17 of their bulletin, they depict a chamise stand before and after treatment.

Likewise, Sumner (1935:294) recommends clearing strips through dense brush, and he further proposes that "where labor and expense are not prohibitive . . . lanes cleared through brush be disked each season. This treatment, by breaking up the hardened soil and checking the growth of certain grasses which tend to grow up fast and choke out other more desirable weed types, greatly stimulates the growth of weeds producing quail food."

Although it is true that quail habitat can be improved by such treatment, Sumner properly suggests the practical limitation of this pattern of management when he alludes to the "labor and expense" involved. Most land operators, private or governmental, can scarcely afford this intensity of treatment on large areas of chaparral solely for the purpose of quail management.

On a more practical scale, strips of chaparral may be cleared as fire-

Figure 69. The Fred Canyon fuel break in San Diego County creates open areas in the chaparral where food-bearing forbs can grow. R. Wakimoto photo.

breaks, and the quail thereby profit incidentally. In the Cleveland National Forest, for example, firebreaks cleared along ridgetops in chaparral offer feeding areas for quail, deer, and other wildlife (Fig. 69). On Water Company lands in the Berkeley Hills where Genelly and Raitt conducted their quail studies, the birds regularly fed along firebreaks and fire trails opened through the dense stands of coyote bush and poison oak.

Where brush stands are opened up for purposes of improving livestock grazing, the effect on quail depends on the extent of brush removal and on the intensity of grazing that follows. As noted in the last chapter, large-scale brush clearance projects are often deleterious to quail habitat. On the other hand, creation of small openings by patch burning or clearing may increase carrying capacity for quail if the ensuing forb growth is not cleanly grazed off. Biswell et al. (1952) studied the effect of patch burning on wildlife populations on a sheep ranch in Lake County. In unbroken chaparral they found 100 quail to the square mile, whereas after treatment the same area supported 250 quail per square mile. The cost of treatment was charged to pasture improvement and reduced fire hazard.

In forest stands, openings can be created by clear-cutting spots or strips, utilizing the logs, and piling the limbs into brush piles as already described. Ray Conway has developed this technique of quail range management to a fine art in the black oak forests of Nevada and Sierra Counties on the west slope of the Sierra Nevada. Openings thus created are planted

Figure 70. Controlled burn on the Keithly Ranch, Lake County. Forbs growing in the burn supply abundant quail food, whereas the dense unburned chamise chaparral (background) is foodless. H. Biswell photo.

to clover, and the surface disked or churned by caterpillar tractor, so that quail food production is actively stimulated in the openings. Conway grazes these properties moderately, thus producing a crop of beef along with the crop of quail. Multiple-use management is feasible if done with skill and judgement.

ARTIFICIAL FEEDING

Artificial feeding of grain to augment quail numbers is a technique of limited usefulness in quail management on open rangeland. The principal problems concern *reliability* of the feeding program and *cost*. If a quail population is supported by artificial feed, it is essential that the feed be available at all times, otherwise a dependent covey of quail can be thrown into confusion and disorder if the food supply fails even temporarily.

The Dune Lakes Club along the coast in San Luis Obispo County fed cracked corn in enormous amounts—6 to 12 sacks per week (500 sacks per season)—on an area of 450 acres. This heavy feeding program, coupled with intensive predator control, elimination of all grazing, and other management measures, produced a temporary population approaching 3,000 birds, or over 6 quail per acre (Glading et al., 1945). Such intensity of management is exorbitantly expensive and not practical in most situations.

At Shandon, McMillan feeds barley to the quail living in and about his yard and thereby maintains a home covey of 800 to 1,400 birds (see Fig. 82). Approximately 30 pounds of grain is distributed each evening, which comes to about 2 ounces per quail daily. The attraction of this supplemental feed holds the covey in place, but if feeding is missed for a few consecutive days the concentration starts to dissipate.

Artificial feeding is a practical way of holding a covey of birds in a ranch yard or suburban habitat (see Chap. 15). On extensive areas of rangeland, however, it is not an economical or practical management procedure, not is it necessary if natural food sources are well managed.

THE HUSBANDRY OF QUAIL RANGE

On most ranges, food supply is the key element that must be supplied to sustain a high population of California Quail. Achieving success in food management is a function of constant, critical attention on the part of the landowner or operator. The basic capital improvement that permits proper husbandry of quail range, including particularly quail food supply, is the barbed wire fence, which permits judicious regulation of grazing. But even in properly fenced terrain, the manager must exercise continuous judgement as to the deployment of livestock to safeguard the quail range. Quail management on the ground is in effect a labor of love, dedication, and the application of considerable ecologic skill.

13

SUPPLYING DRINKING WATER

IS WATER NECESSARY?

In the 1920's, there occurred a spirited debate between two highly qualified naturalists as to whether arid land quail require free drinking water. Joseph Grinnell (1927:528) characterized drinking water as "a critical factor in the existence of southwestern game birds." He was speaking of both Gambel Quail and California Quail in semi-desert situations. Charles Vorhies (1928) acknowledged that quails drank regularly when water was available, but cited many populations of Gambel Quail in Arizona that seemed to thrive and even to reproduce in areas devoid of known drinking sources. Gorsuch (1934), who studied Gambel Quail near Tucson, supported the view of Vorhies (his major professor) that drinking water was an amenity but not a necessity.

Looking back on this argument retrospectively, we realize that both viewpoints are partially correct. Quail can survive desert conditions without drinking when: (1) succulent insects and/or green foods are available; or (2) cool weather reduces water requirements. Normally, however, arid land quail require drinking water to survive periods of sustained heat and drought.

As regards the water requirements of Gambel Quail, Gullion (1956b: 44) states:

> Quail frequently occur many miles from known water, when the temperatures are as high as they get. However, there are two or three facts that must be remembered. First, this business of "known water" is tricky. Oftentimes the nearest "known water" is a stock tank or a fairly reliable spring, whereas actually there may be water in mine tunnels or shafts, hidden seeps, or merely moist soil, much

nearer than you know about. Secondly . . . a fair percentage of quail don't water every day under warm daily temperatures, and under cooler temperatures, below 90°F., many of the birds don't pay much attention to drinking water. Under these cooler conditions, quail could wander quite some distance from water for one reason or another.

Thirdly, there are other sources of moisture on the desert besides water from seeps and springs. Insects, when abundant, apparently provide sufficient moisture for a few quail, and several desert plants provide water-of-succulence, from fleshy leaves and berries. Notable among these are the sandmats or spurges. In some areas the abundance of the spurges permits fairly large numbers of quail to range independently of free water sources. Other sources of moisture are mistletoe and desert-thorn, with their fleshy leaves and periodic berry crops. However, we must still remember that the largest quail populations on the desert ranges are produced in areas where water can be obtained in the free state. It is these large populations that we are trying to produce when we install guzzlers.

In another paper, Gullion (1960:528) elaborates:

On Nevada desert areas the critical period during which water may be an important factor limiting quail distribution extends from about the middle of June to mid-July. This is the period when average daily maximum temperatures reach or exceed 100°F (38°C), after most of the succulent spring annuals have withered, and prior to the onset of the summer thunderstorm activity. Normally, for 11 months of the year the distribution of water is probably unimportant as a factor limiting quail distribution in southern Nevada. However, during that 12th month the proper distribution of water can be extremely important.

It is during this period that the desert upland habitats which furnish adequate food supplies must have a proper distribution of water if Gambel's Quail are to make maximum use of this environment in Nevada. It is for this brief period of time that the development of artificial watering structures in areas of otherwise suitable habitat becomes a practical and useful game management technique.

The Gambel Quail is probably better adapted physiologically to endure water deprivation than the California Quail. It therefore seems likely that the latter species is quite dependent on moisture intake during critical periods of hot dry weather.

MOISTURE REQUIREMENTS OF BROODS

Sumner (1935:193) designed an experiment to test the ability of California Quail to exist and rear young without free drinking water. Two pairs of adult quail kept in a large enclosure near Santa Cruz were deprived of water for a full year, from February 1, 1933 to February 1, 1934. As was expected, the birds showed no evidence of thirst during the winter and spring months. With the advent of summer heat and the drying up of succulent annual vegetation, the birds subsisted by eating green alfalfa leaves growing in one portion of the enclosure. One pair of quail hatched a brood and reared them successfully on the alfalfa, which, however, was completely stripped of leaves by July 20. The birds (adults and young alike) then turned their attention to the coarse leaves of some sunflower plants and existed on this source of moisture until autumn. The second pair of adults brought off a late brood of chicks on July 21, but in the absence of succulence available at ground level the brood perished of thirst. For all practical purposes, therefore, we can conclude that California Quail require water in one form or another to rear young. The need is best met by free drinking water, but, as Sumner demonstrated, broods can be reared on succulence alone when green forbs are abundant.

Grinnell (1927) asserts that nearly all quail nests are situated within 400 yards of available drinking water. On the McMillan ranch this does not appear to be the case. Nesting pairs scatter widely over the countryside in springtime, when green vegetation is at its peak and there is adequate succulence for both breeding adults and chicks. First nestings, and perhaps some second attempts, may succeed in rearing chicks in the absence of free water. In point of fact, because areas near open water are generally overgrazed, the habitat for rearing chicks may be superior at long distances from streams, unfenced springs, or stock watering troughs. By mid-summer, however, greenery diminishes or disappears, and then broods can be reared only in situations where they can drink regularly—meaning the nests must be no more than the cruising radius of young chicks from water (perhaps 400 yards).

Fenced springs or watering devices supply water for quail but exclude livestock. For this reason they offer superior opportunity for quail to rear broods, especially in mid-summer. In the semi-arid foothills, nesting and rearing habitat is far better where livestock grazing is light or absent. In the more mesic coastal and northern ranges, some grazing may improve rearing

Figure 71. A family group of quail drinking at a garden pool. M. Erwin photo.

habitat by thinning dense ground cover, but overgrazing near water is a more common problem than undergrazing.

NORMAL DRINKING HABITS OF CALIFORNIA QUAIL

During hot dry weather, California Quail ordinarily come to water daily. Nesting pairs may water at any time of day, but they tend to favor the morning and evening periods, rather than mid-day. Parents with broods most frequently bring their chicks for the daily drink after the morning feeding period. Michael Erwin spent several days observing water holes from the vantage point of a blind. On one day (June 19, 1973) at a guzzler situated near Shandon, he saw 17 adults and 97 chicks water between 7:15 and 10:05 A.M. No birds came during mid-day hours. Between 6:00 and 9:00 P.M. an additional 40 adults (mostly pairs) watered, but no chicks. However, on other occasions, both adult pairs and family units were observed drinking throughout the day, although in lesser numbers than in the morning.

Up to about the age of 4 weeks, broods are kept fairly well segregated by their parents—that is to say, broods are not ordinarily permitted to mix. If a family of quail arrives at a water source that is already occupied by another group, the new arrivals hold back and await their turn until the occupants depart. Mixed family groups are observed, but they are the exception rather than the rule when the chicks are small.

Brood segregation breaks down as the birds grow older, and by late summer the integrity of individual family units is largely lost in the process of formation of autumn coveys. However, the habit of drinking after the morning meal tends to be retained. The most likely time to encounter quail at a water source, from August to early October, is between 8:00 and 10:00 A.M. When the fall rains begin, the coveys drift away from the watering places and may not drink at all, or only irregularly. Green foods brought on by the rains, and temporary pools left by the rains, supply water needs adequately without visitation to the summer water holes.

As Gullion pointed out, dependable water sources may be critical for only a short period of the year, but without them quail cannot survive in large numbers.

The amount of water consumed by quail is doubtless highly variable from season to season, being inversely proportional to the availability of succulence. Macgregor (1953:160) states: "Each quail utilizes approximately 15 grams of water a day and, depending upon the length of the dry season, up to three quarts a year." I have no comparative data to evaluate this estimate.

DISTANCE TRAVELED TO WATER

Very young broods probably cannot negotiate a trip to water of more than a quarter-mile. As the broods get older and stronger, they can cruise much farther between the water hole and feeding grounds.

In southern California, Macgregor (1953:158) reported the following observation of movements of banded adult California Quail during the summer:

> Checks on movements of banded birds on these areas . . . did give information on daily and seasonal range of quail which is important in determining the maximum space between guzzlers for greatest efficiency. It was found that California quail in semiarid rangelands have a daily radius of movement from one-half to three-quarter miles. When two guzzlers were placed one mile or less apart, there was considerable daily movement by birds between the guzzlers. Hunting season returns on banded birds showed that 48 percent of the birds were shot less than five-tenths mile from the point of banding; 20 percent were shot between five-tenths and 1.4 miles; 23 percent were shot between 1.4 and 2.4 miles; nine percent were shot at distances over 2.5 miles. The maximum movement recorded for this area was four miles.

Optimum quail range should probably have water sources spaced a mile apart, or less.

NEED FOR COVER NEAR THE WATER SOURCE

When quail converge on a source of water during the dry season, they are particularly vulnerable to predation. It is important, therefore, that escape cover be available in the close vicinity of the water to permit quail to drink with reasonable safety.

Livestock-grazing near water sources imposes a major limitation on the usefulness of most water to quail. Streams, springs, and water troughs are gathering places for livestock in dry weather, and more often than not the ground cover for some distance around is closely grazed and all but

Figure 72. Atriplex shrubs and brush piles form a haven for quail at a watering trough near Shandon.

destroyed as quail cover. To drink, the birds are forced to traverse completely open and unprotected terrain, with consequent exposure to predators. Erwin found the remains of 11 quail killed by Cooper Hawks near one exposed spring on the Camatta Ranch. Moreover, overgrazing of terrain within a quarter-mile of the water (the foraging radius for chicks) renders the ground cover inferior for nesting and feeding. One of the simplest and least expensive measures of habitat improvement for quail is the fencing of small patches of cover next to permanent water sources. The same result can be obtained by planting blackberries in the moist soil near water, since this plant is highly resistant to grazing. Temporary cover can be supplied by constructing brush piles near water, but for permanent improvement fenced areas of dense brush or blackberry thickets are preferable. Figures 72 and 73 illustrate quail coverts created near drinking points, making them useful to quail with little sacrifice of livestock forage.

Macgregor (1953:158) compared the productivity of quail populations using water in fenced guzzlers with that of other populations utilizing unfenced natural water sources (Table 14). He states: "These figures combined with field observations indicate that in dry years, when the front country (meaning terrain adjoining natural water sources) is heavily overgrazed by livestock, the nesting cover is destroyed, and poor nesting success and juvenile survival result. Thus, in 1948, there was little nesting cover in the front country and only nine-tenths of a young bird per adult female; whereas, in 1950 the cover was in much better condition, and 4.0 young birds were found for each adult hen." The fenced guzzlers were situated far from natural water; hence the environs were lightly grazed and supplied much better cover for nesting and for brood survival (4.0 to 5.1 young counted per adult hen in each of four years).

The fencing of small cover patches near water is only a partial solution to the problem of supplying optimum breeding habitat, since it offers protection for birds coming to drink but does not improve nesting cover outside the fence. Nevertheless, good results have been obtained by such fencing, and even better results by developing blackberry thickets.

TABLE 14.
Waterhole counts of chicks per adult hen, reported by Macgregor (1953:158) from a study area near Bitterwater, San Benito County

Year	*Grazed sites near available stock water*	*Guzzlers where water was available to quail but not to livestock*
1947	1.8	4.5
1948	.9	4.5
1949	3.0	5.1
1950	4.0	4.0

Figure 73. Even a single *Atriplex polycarpa*, planted near a waterhole and protected from grazing by a fence, permits quail usage, C. Wiley Ranch, Greenfield.

ARTIFICIAL WATERING DEVICES

In addition to the guzzler (to be discussed in the next section), there are many other ways of supplying drinking water for quail where none exists.

True (1933b) describes a device constructed of a 55-gallon galvanized drum to which was attached an automatically fed drinking trough. A number of such water sources were installed on quail refuges in the 1930's, with quite satisfactory results. The San Bernardino National Forest adopted the idea of water drums for quail and experimented with improved designs (Rahm, 1938). Some units were fenced to exclude livestock, but in others the trough was covered with protecting planks which excluded large animals but permitted quail to enter below the planks. Later, an arrangement was perfected with brass nipples attached below the exit pipe. A ball valve in the nipple prevented leakage, and quail quickly learned to peck at the valve, thereby releasing drops of water into the bill. In any event, water supplies in the tanks had to be replenished regularly (32.6 gallons used per 1.4 months) and maintenance became a major problem. Drums were abandoned after development of the guzzler.

Figure 74. Mountain quail drinking at a drum watering device installed on the San Bernardino National Forest. California Department of Fish and Game photo.

Glading et al. (1945) used a combination of watering devices on the "Kettleman Hills Quail Project," including drums, guzzlers, and ¼-inch tubes tapped into the main water lines maintained by the oil companies for fire protection. The steady drip from the tubes not only watered quail and other wildlife but stimulated vigorous growth of *Atriplex* shrubs growing in the moist soil.

Beside his house, McMillan installed a water trough attached to the domestic water supply, with a float valve to regulate flow.

The outlet pipe from a windmill can be tapped to trickle water into a shallow trough fenced to exclude livestock. The main pipe leads to a livestock trough outside the fence.

All of these devices are satisfactory if they are properly maintained.

THE GALLINACEOUS GUZZLER

Glading (1943), together with co-workers in the Department of Fish and Game, developed a self-filling quail watering device which had the enormous advantage of requiring virtually no maintenance. The pilot model was constructed of concrete, poured on site, and consisted of an underground tank of 600-gallon capacity, with a ramp sloping down to the bottom of the tank serving as the access route for quail at whatever level the water stood. The tank was covered with a plank roof (or poured concrete roof) to reduce evaporation and to exclude debris. During the rainy season, water would flow from an oiled apron into a settling basin (which could be cleaned through a trap door), and thence into the tank. The size of the apron could be adjusted according to predicted rainfall for the local area. Each winter the tank would fill, and birds could utilize the water through the hot dry summer and fall. An enclosing fence excluded cattle from the tank and, more importantly, from the oiled apron. Figures 75 and 76 illustrate the guzzler in construction and in operation. Subsequent improvement has come in the form of heavy plastic or fiberglass tanks, prefabricated commercially,

Figure 75. A fiberglass guzzler tank about to be lowered into the excavated hole. California Department of Fish and Game photo.

Figure 76. A well-placed guzzler in use by a large covey of quail. Rain falling on the apron (center) drains into the underground tank to which the birds have access through the double entrance (lower right). California Department of Fish and Game photo.

TABLE 15.
Numbers of quail guzzlers installed by the California Department of Fish and Game through 1974

Region 1		
Lassen County		18
Modoc County		2
Siskiyou County		14
Shasta County		8
	Total Region 1	42
Region 2		
Amador County		5
San Joaquin County		3
Yolo County		2
	Total Region 2	10
Region 3		
Contra Costa County		1
Lake County		6
Monterey County		236
San Benito County		107
San Luis Obispo County		112
	Total Region 3	462
Region 4		
Fresno County		74
Kern County		155
Kings County		6
Mariposa County		2
Stanislaus County		2
Tulare County		3
Tuolumne County		7
	Total Region 4	249
Region 5		
Imperial County		48
Inyo County		20
Los Angeles County		34
Mono County		24
Orange County		32
Riverside County		387
San Bernardino County		333
San Diego County		386
Santa Barbara County		93
Ventura County		37
	Total Region 5	1,394
	Grand Total Statewide	2,157

which greatly reduce the cost of installation. A commercial guzzler is illustrated in Figure 75.

This device proved so satisfactory that it has been widely adopted throughout the arid West for supplying drinking water for quail and other birds. Through 1972, a total of 2,150 guzzlers have been installed in California by the Department of Fish and Game (Table 15), and hundreds of others have been privately constructed by landowners.

A major advantage of the guzzler over dependence on natural water is that guzzlers can be installed in areas far from watering points for livestock, where natural vegetation is grazed little or not at all. Choice of site for installing a guzzler is very important. Adding water to the environment will only benefit quail if the area has adequate food and cover of types required by the birds. Generally speaking, these requirements are best met in arid or semi-arid range lands; hence the concentration of guzzler construction in the southern counties of California (Table 15). On the more northerly ranges of the species, water is not often a limiting factor to quail occupancy.

If quail already exist within 2 or 3 miles of a new guzzler, they usually will find it and colonize the site. Glading et al. (1945) liberated quail at watering points constructed on the Kettleman Hills project, but the birds did not stay. The following year, however, all new water sources were discovered by quail dispersing naturally during the breeding season, and continued occupancy was assured.

As stated in Chapter 3, governmental wildlife agencies have focused their efforts in quail habitat improvement on construction of guzzlers. Supplying water for quail is by no means the most crucial need in habitat management, but it is something that a Department of Fish and Game can do. Regulation of grazing to improve food and cover conditions is much more pertinent to California Quail management over the bulk of the species' range. But regulating grazing is in the purview of the land operator, not the fish and game agency. Water development is effective only in those arid and semi-arid habitats where, for one reason or another, grazing is not excessive and food and cover are adequate, but water is lacking. Under those restricted conditions, guzzler construction has proven effective in increasing quail numbers.

14

HUNTING THE CALIFORNIA QUAIL

HUNTING PHILOSOPHY

Among California hunters, quail hunting has long been the *ne plus ultra* of outdoor sport. On a crisp winter morning, the hike over the hills in search of birds is pleasant and invigorating. When a covey is located, there is opportunity for high strategy (and considerable running) to maneuver and scatter the bunch in low cover where the individual birds will stick and where shooting is possible. The actual shooting of flushed birds requires skill and precision, for the California Quail is a fast and tricky flyer. And perhaps the greatest pleasure of all is the work of the well-trained dog in helping to find the covey in the first place, in locating the scattered singles, and in retrieving every bird that is dropped. At the end of such a day, both the hunter and the dog return to the car with a feeling of mutual satisfaction. Half a dozen plump birds ultimately will grace the dinner table as a final premium of the hunt.

What is the impelling attraction that hunters feel toward this sport? The intensity of dedication and the single-minded obsession that seem to grip a hunter's soul during the autumn months is a source of puzzlement to wives and to non-hunting friends. The Spanish philosopher, Jose Ortega y Gasset (1972), wrote a whole book on the subject, tracing man's hunting instinct from the cave dwellers of Altamira to the present. The crusty old southern sportsman, Archibald Rutledge, offered his own explanation in a delightful editorial that appeared in the October, 1936, issue of *Field and*

Stream. And many others have attempted to define the essence of hunting. In fact, I once had a try at it myself (Leopold, 1972).

The fact remains that a considerable segment of the American public participates in sport hunting, and one component of the conservation endeavor is to maintain populations of game species in sufficient abundance to permit a regulated harvest. In California, there currently are about 670,000 licensed hunters, and of these a quarter pursue the California Quail. The bird is hunted as well in five other states and in Baja California. My primary intent in this chapter is to record the objectives and methods of quail hunting as a prime outdoor sport, with the hope of perpetuating the art. Revival of the qualitative values of the hunt will be a strong stimulus to restoration and management of the resource.

HUNTING RIGHTS

Initially, a properly licensed quail hunter must decide where to go hunting. Fortunately, within the present range of the California Quail there are extensive areas of public land available for hunting in the National Forests, on public domain lands administered by the Bureau of Land Management, and on some military reservations. Parts of these public properties support reasonably good quail populations which can be located with systematic exploration.

On private ranchlands, it is of course necessary to arrange for hunting privileges with the landowner. Some ranchers lease their quail hunting rights and consider the funds so derived as part of the ranch income. Exclusive hunting rights are highly advantageous in permitting the hunter to pursue his sport without interference or competition from other hunting parties, but for this advantage he must expect to pay. Other ranchers, particularly in areas far from urban centers, grant permission to hunt without charge, but such opportunities are diminishing, largely because of past abuses by irresponsible hunters. In any event, pre-arrangement is essential to pursue game on private property.

It goes without saying that hunting parties on either public or private land should respect residences, fences and gates, livestock, and other property values.

HUNTING STRATEGY

Once arrangements are made for a hunt, the next step is to find a covey of birds. A skilled quail hunter can survey the terrain and judge by the habitat where a covey may dwell. Not infrequently, quail can be seen foraging in open areas of low cover, or a sentinel cock may be spotted on his vantage point above the resting covey. Even in proper quail range, however, it is not always easy to locate the birds. California Quail are quite vocal and sometimes can be discovered by listening for their call—"cu-ca-cow"—given most frequently during periods of active feeding in early morning or late afternoon. Five minutes spent listening may save hours of walking in initiating a quail hunt. Arrival on the hunting grounds and the period of listening should be done in silence, since coveys that have been hunted previously will often take to cover and cease calling at the sound of human voices or the slamming of a car door.

Sometimes quail can be induced to answer a simulated call—"cu-ca-cow," produced by the hunter. In Chapter 2, it was noted that market hunters employed this trick to locate coveys, using a call made with a split twig and a leaf. Commercial quail calls are available today.

Gilbert W. Colby of Berkeley described a procedure used to locate quail by an old-time hunter in Marin County (pers. comm.): "This man was skilled in simulating the call of a hawk [red-tail presumably], and upon hearing this call the quail in the area would take wing and fly into nearby cover for protection, thus innocently revealing their location. Invariably, when no quail were roused in a ravine by the hawk call, subsequent hunting of the area by man and dogs failed to locate birds. The method was nearly fool-proof and saved much fruitless searching in unoccupied coverts." I have not seen this procedure used.

Failing any other means of locating birds, the hunter must walk through the areas of favorable terrain, hoping to flush a covey from its concealment. It is at this stage that a good hunting dog is invaluable.

THE ROLE OF THE DOG

To enjoy quail hunting to the fullest, a trained bird dog is almost essential. Quail hunting without a dog is like dancing without music. It is true that some expert quail hunters do not use dogs, and that many dogs taken into the field are useless. However, it is generally agreed that quail should not be hunted without a dog that will at least retrieve lost or crippled birds, and to fully enjoy the sport it is preferable to use dogs of the pointing breeds. Setters and pointers, when properly trained, are excellent retrievers, and to see an eager bird dog, racing at full speed, suddenly scent hidden game and slide into a keen, staunch point, is a sight to be remembered long after the shooting is forgotten.

Assuming that such dogs are to be used, their main role in the hunt is to locate game. In areas of extensive cover and rough terrain, it is sometimes very difficult to find the winter coveys, even where there is an adequate population. After the first fall rains, California Quail can get along without water and are able to range widely and occupy areas remote from where they are found earlier in the dry season. Hawks that prey on quail are more common in fall and winter, and their activities tend to make the coveys seek the safest refuge and to be more quiet and furtive than at other times of the year. In quail season, during November, December, and January, the most enthusiastic and wide-ranging sportsman may complete what he thinks is a thorough search of an area and find no birds, whereas an experienced bird dog would have no trouble finding coveys.

At the beginning of a hunt, a good quail dog ranges freely and indicates

Figure 77. Ray Conway moving in to flush a quail whose presence in the brush pile is signified by the pointing dog. R. Coppock photo.

when birds are scented. This gives the sportsman an opportunity to maneuver the birds into whatever position he thinks will make the best hunting. After the strategy of the hunt is decided upon, the quail are flushed and the course of their flight marked. It is often helpful to have one member of the hunting party posted on high ground to mark the place where the covey alights. This precaution can save a great deal of time and fruitless searching.

A covey of California Quail will often make its initial flight in a tight group, and if not pursued quickly the covey will reassemble on the ground and swiftly run away. A second or even a third flush is required before the birds will scatter and hide individually. In wet weather when the ground cover is heavy with water, quail are loath to hide but will run again and again, offering few good shooting opportunities. Nor will they "stick" well in chaparral where the ground beneath the canopy is open (as under manzanita, chamise, or Scotch broom). In any weather, brush piles and blackberry thickets are advantageous in inducing birds to take cover and await the arrival of the hunting party. In approaching a covey for the initial flush, it is well to attempt to drive the birds away from chaparral or dense woods and toward open country with low cover where both shooting and dog work may be optimized.

After the birds are scattered and stuck, the real sport begins. California Quail, while strong and fast on the wing, are not capable of sustained flight. When forced to fly and followed closely, their ultimate mode of escape is to scatter and hide. Each individual, or perhaps a small group, will choose a particular hiding spot where, by remaining absolutely motionless and with

Figure 78. Retrieving downed game is one of the most important functions of a well-trained bird dog. R. Coppock photo.

their protective coloration, they are safe from the inexperienced hunter as well as from natural enemies. The ability of the species to "vanish" in this way is almost miraculous. Many a sportsman, after flushing a sizeable covey, has abandoned the hunt in failure and despair when he was almost stepping on quail that were hiding all around him.

After the birds have scattered and taken cover, a good dog can locate and point them with precision. This is the climax of the hunt, where all the skill and finesse of bird hunting can be brought into play. Here is where the bird dog must be able to "do it all"—locate the hiding birds; point them accurately and staunchly; and retrieve all birds dropped, including those that may be winged and able to run and hide in dense cover.

It is in this type of hunting, where the birds burst from cover and are taken over a pointing dog in clean wing shooting, that the maximum in sport and recreation is realized with a minimum of loss and damage to the game population. In view of what must be done to restore and preserve the California Quail, and in consideration of its rare sporting qualities, it seems only proper and consistent to require the highest type of sportsmanship of those who hunt this fine game bird. The use of good dogs is a concomitant of sportsmanship in quail hunting.

REASSEMBLY OF HUNTED COVEYS

After a covey has been scattered by hunting, it is an excellent idea, from both the standpoint of the birds and the hunter, to leave the covey unmolested for a few days so that the surviving birds can reassemble. Coveys hunted persistently day after day become completely fragmented and temporarily seem to lose their natural inclination to gather together. Scattered groups of a few birds each, driven from their familiar home range, are probably more susceptible to predation and other forms of loss than organized covies. Likewise, of course, fragmented small groups of birds are almost impossible to hunt satisfactorily. In terms of the shooting obtained, it is not worth the time to pursue and finally scatter a band of half-a-dozen birds.

RECORDING THE RESULTS OF QUAIL HUNTING

One way in which a hunter can increase his pleasure in hunting quail is to keep a journal of his observations and experiences. The number and size of coveys encountered is an interesting statistic to compare from year to year or from one area to another. Likewise, recording the sex and age of each bird taken (see p. 104 for aging criterion) becomes highly significant in appraising the success of the current breeding season and in comparing seasons. Generally, hunting success is high in years when a good crop of young is produced.

A hunter who takes pride in his shooting may want to record the number of shells expended per bird killed. An average of less than two shells per bird represents good shooting, and this usually means a minimum loss of cripples. Consciousness of shells expended during the hunt tends to discourage taking long shots with concomitant high crippling losses.

There has come to my attention a remarkable hunting journal kept by the six members of the Tomales Point Gun Club that hunted quail on leased land in Marin County from 1927 to 1941. The document not only has historical and scientific value but must have been a source of continuing pleasure and interest to the members of the club.

SEASON AND BAG LIMITS

Each state has a Fish and Game Commission (or equivalent) that sets annual hunting regulations for quail. In general, an attempt is made to liberalize or tighten regulations in response to advance information on the status of the quail crop each year. When applied to regions of a state, or in some cases to whole states, this is at best a crude tool for regulating the kill. Nevertheless, there are some general precepts that can be applied in considering the choice of hunting regulations for any given year.

When the basic breeding stock is known to be at low ebb, following a severe winter die-off in northern ranges or a severe drought in arid ranges, there properly should be a curtailment of hunting to accelerate recovery. In my opinion, the initial restriction to reduce the kill should be a reduction in the daily bag limit. With a permissible bag of six birds, or even less, a hunter and his dog can still enjoy a pleasurable day afield. Only if the population level is at an extremely low level should the hunting season be shortened. This step is more restrictive of recreational opportunity and is rarely necessary to protect the breeding stock. The objective of managing a quail population as a hunting resource is to supply recreation rather than meat on the table, so reduction of days afield should be adopted only when really required to protect the breeding stock.

The optimum period for hunting is restricted by natural biological events in the quail population. In arid regions, the birds are concentrated around water sources in early autumn, and they scatter over the range only after the first rains fall. Hunting of waterhole concentrations is disruptive of populations. Moreover, the weather is usually hot, dog work is poor, and the sport lacks quality. Opening of the season should be made to correspond with the anticipated beginning of the fall rains. This is less of a problem, of course, in northern California and Great Basin ranges than in southern California or Baja California.

At the other end of the hunting period, in mid-winter, the birds are under maximum stress from the weather and shortage of food, and hunting may be an unwarranted disruption of the incipient breeding stock. This is particularly true of populations in northerly latitudes, subject to winter snows.

The optimum hunting period in more northerly ranges falls generally in the months of October-November-December, while in the arid reaches of the southern half of California and Baja California the best period is in the months of November-December-January. Within these blocks of time, seasons may be set according to the status of the populations in a given year. A normal hunting period of three months is a desirable objective.

As regards bag limit, it is my contention that a limit of 8 birds is com-

Figure 79. A bag of California Quail. C. W. Schwartz photo.

pletely adequate for a satisfactory hunt and should be adopted as the norm. In years of poor production, the bag may be reduced to 6 or even less to protect the depleted breeding stock. However, these refinements in regulating hunting may be unnecessary, since hunters tend to be easily discouraged and to cease hunting when birds are scarce. Gallizioli (pers. comm.) feels that the hunting kill of Gambel Quail in Arizona is self regulating, depending largely on quail density. Even in years of quail abundance, relatively few hunters kill the allowable daily bag.

As noted above, these modifications of general hunting regulations do not assure a proper allocation of the kill locally. On private lands, at least, there is opportunity to apply much more refined criteria for regulating and distributing the kill. The landowner who assumes husbandry of his quail population can dictate precisely how many birds may be removed from each covey and in the aggregate from his standing crop.

There would be much stronger incentive for landowners to manage quail and their habitat if the law were modified to foster and reward private initiative in wildlife restoration. As presently constituted, the hunting regulations do not differentiate between a landowner who manages his quail and another who wantonly or thoughtlessly destroys quail habitat and who supports no quail at all. New legislation is needed which provides encouragement to manage and harvest wildlife resources in accordance with good principles of animal husbandry. Such legislation has been introduced in the California legislature but to date it has not been enacted.

THE HUNTING KILL OF QUAIL IN CALIFORNIA

The California Department of Fish and Game regularly gathers statistics on hunting in the state by sending questionnaires to a random sample of license buyers. In January, 1974, questionnaires were sent to 22,927 hunters of the 668,500 who purchased licenses to hunt during the 1973 season—a 3.4 percent sample. Approximately 40 percent of the hunters who were contacted submitted replies. On the basis of this sample, and a similar one compiled the previous year, the Department offers the following report on quail hunting in California:

Quail	*1972*	*1973*
Statewide bag	1,334,800	1,774,300
Hunters reported	156,400	173,700
Average seasonal bag	8.5	10.2
Total days hunted	674,100	783,800
Average days hunted	4.3	4.5

This record indicates that approximately one California hunter in four does some quail hunting. During the two years cited above, he hunted an average of about 4.5 days and killed between 8 and 10 quail, for an average daily bag of 2 birds. This is a somewhat lower kill figure than the long-term average, but 1972, it will be remembered, was a poor quail year in the southern half of the state.

The quail kill was surprisingly uniformly distributed over the state in the dry year 1972, but showed a preponderance of take in the more southerly areas in the wet year 1973:

Department of Fish and Game Region (north to south)	*Percentage of statewide quail kill*	
	1972	*1973*
Region 1	16	10
Region 2	19	14
Region 3	20	21
Region 4	22	24
Region 5	23	31
	100	100

These data correlate closely with the records of quail age ratios reported by Erwin in Appendix C. In a given year, the areas of highest production of young quail will yield the highest bags to hunters.

PREDICTING THE QUAIL CROP FROM SPRING CALL COUNTS

As noted in Chapter 9, male California Quail display marked aggressiveness and reproductive enthusiasm in years destined to produce a good crop of young birds. In years of poor reproduction, aggressive behavior is minimal and ceases early in summer. One of the easily observable criteria of aggressiveness is vocalization (''cow'' calling) by unmated males. Systematic recording of the amount of calling heard in a given spring should form a reasonable basis for predicting the forthcoming success of the nesting season and the subsequent hunting season.

Records of vocalization have not been systematically kept for California Quail, but they have for Gambel Quail, as reported by Smith and Gallizioli (1965). Weekly calling counts made at sunrise along pre-determined routes through Gambel Quail range in Arizona, from April 15 to May 20, yielded comparative data that seemed to predict accurately the success of breeding and subsequent hunter success the following shooting season. Table 16 taken from their report shows that predictions averaged within 10% of the subsequent actual kill of quail per man day of hunting, which is accurate enough to predicate hunting regulations for the year. This system doubtless could be used successfully for California Quail in the arid portions of their range, where substantial year-to-year differences occur in reproductive success. The applicability of the method is questionable in the more humid ranges where annual productivity is relatively stable.

TABLE 16.
Comparison of spring predictions of fall hunt success and actual kill per man-day on Gambel Quail ranges in Arizona
(After Smith and Gallizioli, 1965:812)

		Quail bagged per man-day		
Study area	*Year*	*Predicted**	*Actual*†	*Percent difference*
Oracle Junction	1963	4.53	4.83	+ 6.6
	1964	3.52	2.91	−17.3
Pinnacle Peak	1963	1.65	1.70	+ 3.0
	1964	1.34	1.60	+19.4
Cave Creek	1963	1.63	1.82	+11.7
	1964	1.34	1.35	+ 0.7
□ Average deviation				9.8

*In early May from call-count data.
†From check station data, October.

ALLOWABLE HARVEST

In an average year of reasonably good production, what is the allowable harvest? Data on hunting intensity of the California Quail are meager. Glading and Saarni (1944) reported harvest rates of 18.6 percent to 26.5 percent of the fall population on a public shooting area in Madera County. When unrecovered cripples were added to the bag, the removal rates were between 24 and 38 percent (see Table 17). Emlen and Glading (1945:49) give 25 percent harvest as a tested level of safe removal. On the McMillan ranch near Shandon, where the birds are shot very conservatively with no attempt to maximize harvest, annual hunting removal varies from 10 to 30 percent of the fall population with a mean of about 15 percent. Hunting pressure and percentage of birds bagged on this area has been lowered in recent years, even though the total population has been raised through habitat management.

As noted in Chapter 10, Raitt and Genelly (1964) removed 70 percent of their study population one year and 41 percent the next year without reducing the population level in the year following. However, they concluded that the losses were compensated in part by an influx of breeders from adjoining areas, and they made no assertion that hunting removals of this magnitude could be sustained over large areas.

On properties between Grass Valley and Marysville, intensively managed by Ray Conway for quail hunting, the range of hunting removal varies from 20 to 40 percent, averaging about 30 percent. On one of the better pieces of Conway's quail range ("320 East"), the harvest in 1973–74 was 154 birds out of an estimated population of 653 on 320 acres, for a 25 percent yield. Most of the shooting took place on 150 acres for a kill of a bird per acre. In 1971–72 the bag on the same property was 228 birds. Conway considers that the kill could be even higher without adverse effect on the breeding stock.

More is known about Bobwhite hunting on the intensively managed quail plantations of the Southeast. Rosene (1969:206) states: "Records from the South indicate that it is safe to remove about 45 percent of the (fall) population by hunting. This would apply wherever loss in winter due to weather is negligible. A 45 percent removal means birds brought to bag, including loss by crippling. The same percentage would apply in the North but here loss due to weather after the hunting season closes must be anticipated and added to the loss by hunting so that the sum does not exceed 45 percent."

Annual productivity in southern Bobwhite populations is higher than productivity in most California Quail populations. Fall age ratios in Bobwhites average 300 young:100 adults, whereas in California Quail the ratio is more nearly 150 young: 100 adults, with enormous fluctuations between years. The high turnover rate and higher production of young in Bobwhites suggest that they can withstand a higher average harvest than California Quail on semi-arid ranges. I would suggest that 30 to 40 percent should be considered the normal range of allowable harvest for California Quail.

In the northern parts of the California Quail ranges, turnover rate in the population is as high as in Bobwhites. In a seven-year period, Savage (1974) found the average age ratio in Modoc County to be 353 young:100 adults. Anthony (1970a) reported a two-year sample of birds taken in eastern Washington to run 236 young:100 adults. Despite the higher rate of productivity in these populations, I doubt that allowable harvest rate should exceed 30 to 40 percent because of the probability of winter losses.

DOES HUNTING ENDANGER QUAIL POPULATIONS?

Excessive hunting can temporarily reduce the average level of a quail population, but hunting at moderate levels probably has little effect on year-to-year density. As noted in Chapter 8, there is a normal annual cycle in a quail population which leads to production of a substantial surplus of birds in the autumn that will disappear in one way or another during the ensuing year. Moderate hunting is not additive to normal mortality but rather substitutes for it. Unhunted populations of California Quail studied by Emlen (1940), Genelly (1955), and Raitt and Genelly (1964) display the same sort of annual cycle as the hunted populations studied by Glading and Saarni (1944) and Anthony (1970a).

The study conducted by Glading and Saarni is instructive, since these investigators were directly measuring the effect of hunting on population levels. Table 17 is taken from their report, and the data suggest that over a

TABLE 17.
Summary of census data and of bag and cripple losses on experimental hunting areas, San Joaquin Experimental Range, 1938–1942
(After Glading and Saarni, 1944:75)

Hunting area	*1938–39*	*1939–40*	*1940–41*	*1941–42*
Pre-season (November) census	513	245	278	349
Total bag	136	62	52	84
Unrecovered cripples	57	26	14	25
Total bagged and crippled	193	88	66	109
Percent removal by hunting	38%	36%	24%	31%
Post-season (March) census	214	141	167	184
Percent population increase, March to November	15%	97%	109%	74% (av.)
Check area (unhunted)				
Pre-season (November) census	349	288	256	325
Post-season (March) census	257	185	216	202
Percent population increase, March to November	12%	38%	50%	33% (av.)

four-year period the trends in post-season (March) populations of California Quail were essentially alike in hunted and unhunted populations in similar habitats.

There is the classic case, often cited, of the status of Bobwhites in Ohio and Illinois. No quail hunting has been allowed in Ohio for over 50 years—the Bobwhite is on the songbird list—but populations there are no higher than in adjoining parts of Illinois where moderate hunting seasons are declared each autumn. Quail cannot be stockpiled by statute.

Errington and Hamerstrom (1935) studied the post-season survival of Bobwhites on hunted and unhunted ranges in Iowa. On the hunted area, 2,100 birds surviving the hunting season and going into the winter decreased to 1,882 in spring, for a 10.3 percent winter loss. On the unhunted area, 1,132 birds going into the winter dropped to 770 in spring, for a 31.9 percent loss. They concluded that hunting had removed some of the "vulnerable surplus" of birds, thereby improving the chances for survival of the remainder.

In New Mexico, a study was conducted for 8 consecutive years of two populations of Scaled Quail, one on a heavily hunted area and the other on a comparable check area that was completely protected from hunting (Campbell et al., 1973). For the full 8-year period, the populations of the two areas remained essentially equal. The mean March population on the hunted area was 21.9 quail per square mile, following the previous autumn removal by hunting of 17.1 quail per square mile. On the unhunted area, the mean March population was 29.2 quail per square mile, without any hunting removal. The autumn populations of the two areas following the subsequent breeding season were near equality, and the authors state (p. 26): ''It appears that proportionately more young quail were added to the treatment [hunted] area than to the control area, thus accounting for the relative increase of the former.'' This case well illustrates the principle of cropping a population on a basis of sustained yield, leading to a compensatory increase in annual production. The same trend is evident in the California Quail data gathered by Glading and Saarni (1944) and presented in Table 17. Average population increase from March to the following November was 74 percent in the hunted population, and 33 percent in the unhunted population.

Gallizioli (1965:3) summarizes data from 12 years of study of Gambel Quail populations on two areas in central Arizona, one subject to annual hunting, the other closed to all hunting for 10 years and then opened. He states:

> The ''control'' area was kept closed to hunting from 1949 through 1959. Quail numbers there went up and down about the same as on the adjacent area which was hunted each year. However, the final, undisputable proof that quail cannot be ''stockpiled'' by protection from hunting was obtained in 1960 when this area that had been closed for 10 years was again opened to hunting. Checking station data showed that hunter success was actually lower on the ''control'' areas than on the adjacent area which had been hunted annually. The difference in success may not be significant. What is important, however, is that hunting success was no higher despite the protection afforded by 10 years of no hunting.

Gallizioli's data on hunting removal in relation to quail population density illustrate another important point—namely, that the *percentage* of the population removed by hunting is high (20 to 30 percent) in years of high quail density and low (4 to 7 percent) in years of quail scarcity (see Fig. 80). Hunting success apparently is subject to the law of diminishing returns.

The data cited above indicate that a reasonable crop can be taken from a quail population by hunting without jeopardizing the future breeding stock. But this is not to be construed as an argument for ever more lenient and unrestrictive hunting regulations. I return to the point made earlier in this chapter that the objective of a hunting experience is not a maximum bag but rather a pleasurable day afield. With quail populations in California still decreasing, it is important to optimize the recreational yield of each bird taken. Both regulating the kill and the training and education of quail hunters

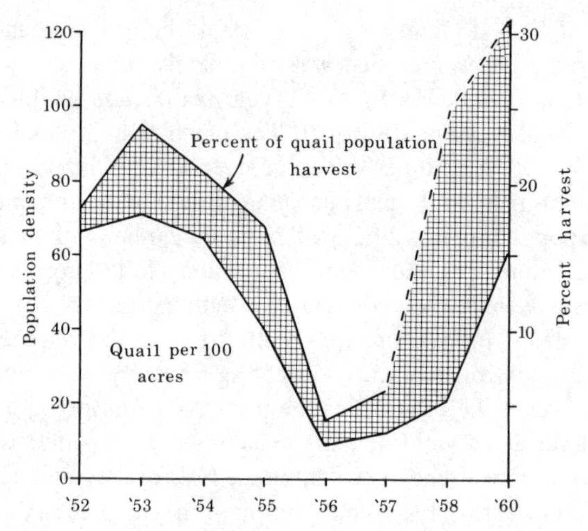

Figure 80. Percent harvest by public hunting of a Gambel Quail population near Oracle Junction, Arizona. (After Gallizioli, 1965.)

should stress the ethics and mores of the sport, not the magnitude of the yield.

To sum up, in extensive areas of California Quail habitat, where hunting can be allowed without compromising other social values, it is a legitimate and justifiable use of a crop produced annually. Moreover, to its faithful "aficionados" it is one of the finest of outdoor sports.

15

BACKYARD QUAIL

QUAIL FOR PLEASURE

There is no more delightful bird to have around the yard than the California Quail. Its attractive appearance and mellow calls endear the quail to nearly everyone, and many householders living in suburban or rural settings would like to attract a covey and keep it nearby, with no thought of hunting. Fortunately, the species adapts well and quickly to co-existence with people. It is entirely practical to maintain a backyard covey of California Quail.

The habitat requirements of quail in a suburban situation are identical with those on the open range. That is to say, the same types of cover are needed for escape, loafing, roosting, and nesting. There must be year-round food available for the adults and special food resources for rearing young. And quail in the yard, like quail anywhere else, require drinking water during periods of heat and drought. Planning a quail management program for a backyard has all the same components as planning for a whole ranch, but on a smaller scale.

COVER

Shrub cover in one form or another is essential. Moreover, the shrubs must be dense at ground level to furnish real protection from predators and a place where the covey can loaf in mid-day. Many suburbs are intersected by streams, ravines, or slopes too steep for house construction or intensive gardening. An acre or two of rough ground, heavily grown with woody vegetation, may suffice as the "core area" of a covey range. In the heart of Berkeley, where building lots are narrow and houses are built close together, I know of a number of wooded and brushy canyons that hold quail. In the center of residential Reno, quail are abundant along the steep, brushy banks of the Truckee River (Fig. 81).

Natural cover in covey core areas can be augmented and improved by planting. For example, in a moist situation, as along a stream bank or a canyon bottom, blackberry will grow profusely if a few root stocks are planted. A tangle of blackberry vines 20 feet in diameter and 6 feet tall is the absolute acme of quail shelter, serving as cover for escape, loafing, and roosting, all in one. Toyon, pyracantha, holly, and other decorative berry plants may form quite satisfactory thickets. In more arid situations, manzanita, *Atriplex,* or juniper can meet the cover requirement. I know of a ranch yard east of Cholame where a short row of dense tamarisk (salt cedar) holds a small covey of birds. In any situation, a covey range must be built on a core of thick cover of one sort or another. Without such a core, marginal types of semi-open cover which may be used intermittently by quail will not suffice to support a flock of birds. There must be a secure "home base."

Nesting cover is one of the more difficult requirements to be met. In open country, quail pairs scatter widely, distributing their nests thinly over the terrain, which is a form of insurance against predation. Such a procedure is not usually possible in suburbs, where a few vacant lots may constitute all the available open nesting cover. As a result of limited nesting sites, I suspect that productivity is low in suburban quail populations. William Molini of the Nevada Department of Fish and Game has submitted to me some data collected in 1973 and 1974 indicating that the proportion of young quail in a sample of birds trapped within Reno was about half that of samples

Figure 81. Urban quail habitat along the Truckee River in the heart of Reno, Nevada.

taken by hunters in nearby rural areas. Domestic pets and other suburban predators such as raccoons, opossums, and rats probably destroy many nests clustered in the few situations suitable for nest location. Still, experience shows that city quail produce enough young to maintain populations, despite nesting adversities.

FOOD

One requirement easily met is supplying food for quail during the periods of the year when the birds are in coveys. California Quail come readily to feeding stations where small grains or mixed bird feed is offered. Feed may be scattered on a patch of bare ground adjoining shrub cover, or it may be offered on an elevated feeding tray which is a bit more secure from the attacks of prowling cats. There is one site in Berkeley where quail regularly visit a tray mounted on the railing of a deck 20 feet above the ground. Ordinarily, however, quail will more quickly discover and accept food scattered on or near ground level.

Although dry seeds are a welcome food supplement used readily by California Quail, they are not of themselves an adequate year-long diet. As explained in Chapter 8, sprouting green plants—preferably legumes—appear to be sought by the birds in winter and spring, and insects plus weed seeds are required by small chicks. Open ground on vacant lots or south slopes is the best site for quail to find sprouting annuals such as bur clover, sweet clover, lupine, or other palatable greens, and the same areas may at a later date in the spring furnish insects and seeds for the young. In the absence of wild greens, quail often turn their attention to sprouting vegetables and

Figure 82. Quail are quick to take advantage of a handout of grain.

Figure 83. California Quail using a backyard bird feeder at Point Reyes, Marin County. G. Heath photo.

flowers in garden plots, to the distress of the home gardener. Voracious quail can deal severely with emerging radishes, beans, sweet peas, or other garden annuals, and complaints of such destruction in Reno have induced the Nevada Game Department to trap large numbers of birds which are then used for stocking outlying parts of the state. Backyard quail are not universally appreciated by ardent gardeners!

WATER

Drinking water is rarely a factor in the welfare of suburban quail. Lawns and gardens are watered regularly during hot, dry periods, and the birds can drink quite adequately from the droplets left on vegetation. Yet it is well to maintain a permanent drinking pool where quail (including young chicks) are assured of water when needed. A bird bath mounted on a pedestal and situated near shrub cover is quite suitable for quail except during the two-week period when newly-hatched chicks are unable to fly. A ground-level bird bath is preferable to meet the needs of young quail. Such a pool can be filled with a garden hose, or, better yet, it can be connected to the domestic water supply for automatic filling (see Fig. 71).

CATS AND DOGS

Perhaps the most difficult aspect of maintaining backyard quail is devising protection from the constant molestation by domestic pets. Dogs rarely catch quail large enough to fly, but they do break up nests and create disturbances of birds going about their daily lives. Cats, on the other hand, not only molest quail, but skillful individuals capture them frequently. Woven wire fencing can be used to exclude dogs from critical areas of quail habitats, but cats are difficult to fence out. Feline pets that are fed regularly are not dependent on catching birds for a living, but rather they hunt for pleasure and avocation. They can afford to spend many happy hours stalking quail and other birds around the yard, and hence they are much more dangerous predators than truly feral cats that must hunt for a living and therefore seek small mammals almost exclusively (wild-living cats rarely catch birds).

It is pointless to ask neighbors to confine their cats, nor is the problem solved by belling. The best protection for suburban quail is to maintain dense cover which stray pets cannot easily penetrate. Blackberry tangles are far and away the best coverts for this purpose.

A QUAIL CALL AT DAYLIGHT

The swelling tide of urbanized, mechanized Americans, living like robots in the asphalt jungles of the city, has created a growing sensitivity toward things that are wild and free. Bird watching and bird feeding are increasing in popularity among city dwellers and suburbanites. Among those that have large yards or live next to vacant lots or other undeveloped plots of land, it is a logical extension of enjoying the presence of birds to managing habitat for the encouragement of backyard birds in increasing variety and abundance. Several recent issues of *National Wildlife* (Oct.–Nov. 1973, and Dec.–Jan. 1974) include specific suggestions on developing urban havens for bird-life.

In my judgement, the creation of living space for a covey of California Quail would represent the gold standard of successful backyard management. What more pleasant sound could there be to awaken a jaded suburbanite than the morning call of the quail—"cu-ca-cow"?

EPILOGUE

The quail covey emerged from the end of the *Atriplex* hedge and gathered in a loose group under the roses at the lower end of the yard. The rancher and his wife, sitting on the verandah to catch the warming spring breeze, knew that the birds had finished their evening feeding and would soon fly into the live oaks to roost. In one of the live oaks, the rancher and his boy had constructed an elevated brush pile in the crown of the tree. Heavy limbs from another tree had been hauled up into the oak and wired securely in place to form a dense shelter for the roosting quail. Most of the birds in the covey would finally settle for the night in that artificial shelter, but some always chose to roost alone or in small groups in other leafy trees about the yard.

From the verandah, one had a clear view of the lower reaches of the ranch. Pale gray ribbons of *Atriplex* separated the fields—some now green with well-grown winter wheat, others newly fallowed and dark brown in the fading light. The main watercourse through the ranch was heavily grown to willows and junipers, from which presently could be heard the distant calling of quail from the adjoining covey.

"Cu-ca'-cow, cu-ca'-cow."

One of the cock birds under the roses took a step forward, raised his head, and answered:

"Cu-ca'cow, cu-ca'-cow, cu-ca'-cow."

There came then an answer from the covey on the brushy hill to the south and even more dimly a far-away call from the covey near the old barn to the east.

The cock that just called had been standing close behind a hen, but turned and made a rush at another cock that approached the pair from the rear. The second cock stood his ground, and there ensued a brief but furious fight, the two birds leaping into the air, sparring with feet and bills. It was clear that the season of pairing and nesting was imminent.

The good quail population on the property had not come about by accident. During off seasons, when farming operations allowed, the rancher and his son had planted many gullies and odd corners with *Atriplex* and juniper, and built fences to protect the plantings from his own cows. Quail watering devices were installed in areas far from natural water, and a few roosts had been constructed to serve until the junipers grew up and assumed this function. Now the rancher took pride in carrying a normal population of a bird per acre on the property as a whole, which is good by any standard. In the autumn, he and the boy would invite friends for a few good quail hunts, but most of the year the pleasure was in watching the birds and anticipating their needs.

The covey under the roses began moving toward the roosting area, largely arranged in pairs—a cock following close upon the heels of his chosen hen. The first pair flew into the roost tree with a whir of wings and hopped about on the brush pile, choosing its site to enter. Then came a burst of birds, and there followed ten minutes of settling into the roost for the night, with much clucking and twittering. As dusk fell, the fluttering and vocalization diminished, and finally all was quiet for the night.

The show was over, so the rancher and his wife went into the house and closed the door to the verandah. As the lights went on, a horned owl hooted in the canyon below.

APPENDIX A
QUAIL IN ABORIGINAL CALIFORNIA

BY KAREN M. NISSEN
Department of Anthropology
University of California, Berkeley

Quail were widely utilized by the aborigines of California. In writing of the Pomo, Barrett (1952:98) states: "Perhaps no other kind of bird was more esteemed as a food than the quail. Certainly no other land bird was more used." Special traps and snares were built solely for the purpose of capturing quail (Kroeber, 1925:817). It has even been proposed by Johnson (1972) that quail were introduced on Catalina Island by Indians as a food source. In some areas, the top-knots of quail were used for decoration of baskets and clothing, particularly among the Pomo and Yokuts. The ethnographic literature is not always precise as to the species of quail involved, often citing "quail snares" or "quail nets" without specifying whether the quarry was the California Quail or the Mountain Quail. Both species were utilized, but in most areas it was the California Quail that was available for capture.

The information summarized here was collected for the most part after the turn of the 20th century, when many details of aboriginal practice had been forgotten or lost. Some native groups were exterminated and no records were available. This summary, in short, is doubtless incomplete. Figure 84 depicts the distribution of aboriginal tribes in California.

Figure 84. Distribution of aboriginal tribes in California.

METHODS OF CAPTURE

The methods of obtaining quail varied from the northern part of California to the southern, with the greatest diversity of methods occurring in the central part of the state from the western slopes of the Sierra Nevada to Clear Lake, where the Miwok, Pomo, Maidu, and Patwin tribes lived.

In the central Sierra Nevada area, among the Miwok, Northern Yokuts, and some Mono and Maidu groups, there were professional quail hunters. This seems to be the only area of the state where such specialization was developed. Beals (1933:349) notes that among the Southern Maidu (Nisenan) some hunters made quail capture a profession, and that they traded the birds for acorns and other meat.

One of the most common methods of capture, and one which seems to have been utilized mainly in the northern two-thirds of the state, was a special quail fence built near a spring or along a hillside with nooses set in

Figure 85. Hair noose set for quail in openings left in brush fence.

gaps. Barrett and Gifford (1933:183–184), whose illustration is reproduced in Figure 85, give a fairly detailed explanation of this method of capture for the Miwok:

The most important food birds to the hill Miwok were the Valley Quail, *Lophortyx californica,* and the Mountain Quail, *Oreortyx picta.* They were taken chiefly by means of a brush fence with openings five to ten feet apart, in which were set snares. Such a fence was built in any desired shape, governed by its situation. It was ordinarily about three feet high and composed of vertical stakes driven into the ground, with twigs or brush interwoven. It was usually built near or around a spring. In the latter case a brush roof was placed over it. Further, any other convenient supply of water was covered with fine grass or pine needles.

The snare was made of human hair twisted into a very fine, but strong, thread. . . . [E]ach opening in the fence had a pair of crossed sticks to which the snare was attached, by very lightly tying it to the point at which the sticks crossed, or to one of them, or by looping it over slight notches cut on the two sticks. These were simply means for holding open the loop of the snare, the opposite end being firmly attached to any convenient branch or stake. The slight notches in the side sticks usually had to be renewed after a quail was caught, because of damage done by its struggles. No attempt was made to hang and strangle the bird by means of a spring pole. It found itself caught by the head, wing, or foot and worried itself to death in its endeavor to gain freedom. A quail fence was used year after year with whatever repairs were necessary. The vicinity of a small streamlet, called "Indian spring" (Kolakota), lying 200 or 300 yards up the canyon from the steatite quarry Lotowayaka, near the modern town of Tuolumne, was a favorite place for erecting a quail fence. The exact site of the fence was a steep ridge just northeast of the streamlet. As many as fifty quail a day were caught in spring when migrating to higher altitudes. Another fence on a hill northeast of Tuolumne was said to have been half a mile long.

The Central Miwok in the vicinity of Murphys placed the fence on the west side of a hill in spring and the east side in fall, "because quail go into mountains in spring, into valley in fall'' (Aginsky, 1943:451). Kroeber (1925:410) states that: ''Quail will often follow a low fence rather than fly over it, particularly along their runways. A fine noose and bait at occasional gates usually stopped a bird.''

The Pomo fence had a number of the V shapes in its length, in which were placed either basket or net traps or treadle snares. The fence was normally placed near brush, as the Pomo believed that quail would be less likely to fly when they were driven through brush. The use of a fence in hunting quail extended from the Yokuts and Mono on the south up to the west slope of the Sierra to the Miwok, Nisenan, Atsugewi, Shasta, Wintu, and Tolowa, and down the coast to San Francisco Bay in the area of the Southern Pomo. It is not known if this method was used along the coast south of San Francisco Bay, as we have no specific record of quail capture for the groups which lived there; for the most part, they were extinct by the time ethnographers entered the area.

A variation on the fence with snares was the use of a net placed at the narrow point of a V-shaped fence along the side of a hill. The net was supported by forked sticks and reached to the ground. The Nomlaki or Central Wintun used this method to capture quail, and according to Goldschmidt (1951:401) quail was the most important food bird to this group. Two men would remain at the net while other people spread out over a wide distance and herded the birds by clapping two rocks together. The props supporting the net were pulled out so that it fell and entangled the quail. Twenty-five to 30 birds were often captured in such a drive. At times, nets were said to have been torn down by the great numbers of birds (Goldschmidt, 1951: 402). Loeb (1926:165) mentions the use of a fence and net among the Pomo, and he also notes that black seed was scattered around the mouth and inside the net to attract the birds. The fence and nets belonged to one family and were removed after use and prepared for future hunts. This implies some sort of semi-professional status for quail hunters in the Nomlaki area.

Similar in function to the fence was the use of a long net. Barrett and Gifford (1933:184) describe this hunting device as follows: ''Nets of human hair and, in Caucasian times, of horse hair were stretched across a hillside up which quail habitually walked. There were openings in the net which led into pockets, flanked by sticks to hold them in position. The quail tried to go through the openings and became entangled in the pockets of the net.'' This method was used by northeastern Yokuts and Northern Miwok.

The Pomo used a long net in a different manner. It was pinned down on three sides, and the fourth side was raised slightly above the ground on a trail known to be used by quail. The hunter hid in a brush shelter, holding onto a string which raised the net. When the quail gathered under the net, the string was pulled and the quail became entangled (Barrett, 1952:135).

A device known as a bag net for snaring quail was also widely used in northern California, from the Modoc on the north to the Miwok, Patwin, and Maidu on the south. Kroeber (1932:279) describes the net and its use as follows for the River Patwin: "Quail were taken in a tunnel-like net, some ten yards long, a foot or two across, held open by hoops. Several men 'made medicine' by tapping two sticks together as they gently drove the quail in without alarming them to break into flight. If one bird entered, the others followed, and a concealed hunter leaped to the mouth of the net."

The Nisenan or Southern Maidu may have captured quail in a similar manner. Kroeber (1929:284) notes that they caught quail in a net placed over seed bait along a quail path. Nets were also set to drop over hillside springs to catch birds as they came to water (Beals, 1933:349–350). An unusual feature is said to be the use of seeds soaked in manzanita cider which were supposed to have been used to attract quail. A Nisenan man said that the birds became intoxicated, but Beals believes this to be unlikely, as cider was not fermented aboriginally.

A net draped over cascara bushes where quail roosted was used by the Valley Maidu (Voegelin, 1942:169). The bag net was often used with a frame; the Yurok had the A-frame of a fish net, while the Yuki used a hoop. Foster (1944:163) describes the Yuki net as follows: "A trap consisting of a four foot hoop covered with a net was laid on the ground and raised on one side with a trip. Seeds scattered beneath attracted a covey, and when the concealed hunter deemed he had sufficient number, he released the trip with a string." The Southern Pomo said that men wore pepperwood and other leaf disguises on their heads and drove the quail into the net. Nomland (1938:112) notes that nets were also used by the Bear River people in hunting quail.

The net was replaced in other areas by a basket trap. This device appears to have been especially used in the area of the Pomo, a group well-known for its excellent basketry. The basket trap seems to have been limited to the area of the Bankalachi tribe, north to the Yokuts and Northern Miwok, and west to the area of the Pomo, River Patwin, and the Kato. Among some of the Pomo groups, the basket trap was set at the opening between two low converging fences (Gifford and Kroeber, 1937:174; Essene, 1942:3). The Central Pomo piled brush over the trap to keep other birds away; quail were driven into the basket trap. The converging wing fences were about two feet high and sixty feet long. Barrett (1952:134) notes that, among the Pomo, men made the basket used for capturing quail out of unpeeled branches. This is unusual, as women were usually the basket makers. The traps would be up to 24 feet in length, made of separate sections from 5 to 6 feet long (Fig. 86). These traps were set in the fences at angles, the fence forming a number of V-shapes where it angled off into the brush. The traps were only six or seven inches in diameter, so quail were unable to retreat once they had entered the trap. Other members of the covey would follow the leading quail into the basket. When the birds would attempt to escape and stick their heads out through the weave of the basket, the hunters would

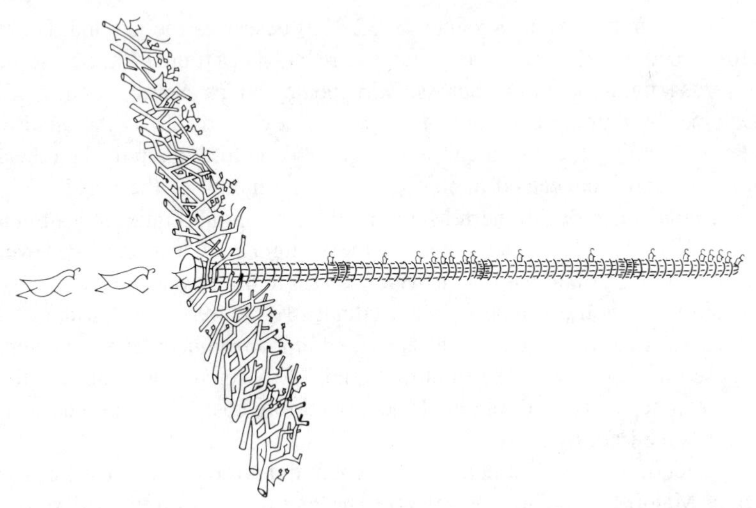

Figure 86. Diagram of a Pomo basket-trap for quail. The brush fence (left) served to guide driven quail into the tube-like basket. Lowie Museum of Anthropology photo.

snap their heads or stun them. One Southern Pomo informant mentioned that quail were driven on rainy days, presumably to take advantage of the lesser inclination of quail to fly when their plumage was wet (Gifford and Kroeber, 1937:174; Barrett, 1952:135). Loeb (1926:165) describes the use of the basket trap as follows: "In the summertime long baskets with funnel shaped mouths were set out in the open, and a little black seed placed inside."

Sliding noose snares were also used by some groups, especially the Pomo, to capture quail. The snare was similar to that used in the long fences. Beals (1933:349) mentions the use of snares made of hair placed low near the water where the birds came to drink. The snare may have been widely used in California for the capture of quail, but there is no mention of it in most of the ethnographic literature other than for the Southern Maidu, Yuki, Wappo, Pomo, Tulare Lake Yokuts, and the Tolowa at the extreme northwest end of the state.

Faye (1923:39) discusses the Southern Maidu method of capturing quail with a snare: "In summer also they would snare quails, using to that end sliding loops made of women's hair. Little hedges about a foot high were built across the country with openings provided in certain places where the quail rushing through would hang themselves by the neck." This may be slightly different from the Miwok fence with nooses, which was meant to ensnare the birds but not to hang them. The Wappo snared Mountain Quail with bent sapling traps (Driver, 1936:185). The Tolowa of extreme northwestern California and other groups to the north along the southwest Oregon coast caught quail, rabbits, and other game by means of a small noose suspended along runways (Drucker, 1937b:234). Yokuts in the vicinity of Tulare Lake also used snare traps for quail. When Fages was touring in the Yokuts territory in 1775, he remarked upon the large numbers of quail in the area. However, when Gayton visited the area during the 1920's, he was told that "California (Valley) quail (saka'la) were scarce in old days but common now."

The trap used by the Yokuts functioned as follows: "[a] spring trap [was] set up on trails of quail, cottontail rabbits, ducks or geese. A small supple limb was set in the earth if none grew naturally at the spot. To its tip was fastened a noose of string, the cord of which was tied to a horizontal bar pegged about six inches off the ground. Any creature caught in the noose struggled sufficiently to release the horizontal bar, and animal and bowed limb flew upward" (Gayton, 1948:75).

On the east side of the Sierra Nevada, quail traps apparently were unknown in aboriginal times. In writing of the Owens Valley Paiute, Steward (1933) mentions the absence of such devices. Nor were any found in Lovelock Cave, Nevada (Loud and Harrington, 1929). In more recent times, quail were trapped by the Northern Paiute in Surprise Valley, but Kelly (1932) explains that they were introduced by the first white settlers. There were no quail originally in northeastern Modoc County.

A figure-4 boxtrap was apparently used only in the extreme northern and southern sections of the state, being more common in the south. The Tolowa and groups along the Oregon coast used a quail trap made in the form of a "pyramidal 'box' of sticks laid up crib fashion" (Drucker, 1937b: 234; see also Barnett, 1937:164). The boxtrap, often with a figure-4 release, was used by the Luiseno, Desert and Mountain Cahuilla, Cupeno, Serrano, Mountain Diegueno, and the Yuma. According to Drucker (1937a:47), "The 'box' was made of sticks laid up crib fashion, in pyramidal form. A figure-four release was used most commonly, though some informants said a simple twig prop, jerked by a string by a hidden watcher, was used."

Pomo groups used a very complicated device known as the treadle snare for capturing quail. The illustration drawn by Barrett (1952:136) is reproduced in Figure 87. Barrett describes this "Rube Goldberg" type of snare as follows (p. 137):

> The treddle [*sic*] snare was set at one of the openings in the fence in such a manner as to bring the treddle bars directly opposite the opening and parallel to the fence. This means that the arch (a) was set at right angles to the fence and a little to one side of the opening.
>
> One end of each of the treddle bars (d) rested on the cross-bar (b) of the trigger, the other end resting on the ground at the opposite side of the opening in the fence. Thus these several treddle bars were parallel to the fence and formed a little inclined platform right at the opening. The loop of the snare (g) was laid flat on this platform and in instances was held open and in place by the four pegs (c), as was always done in the baited trap. With this arrangement of these elements it was impossible for

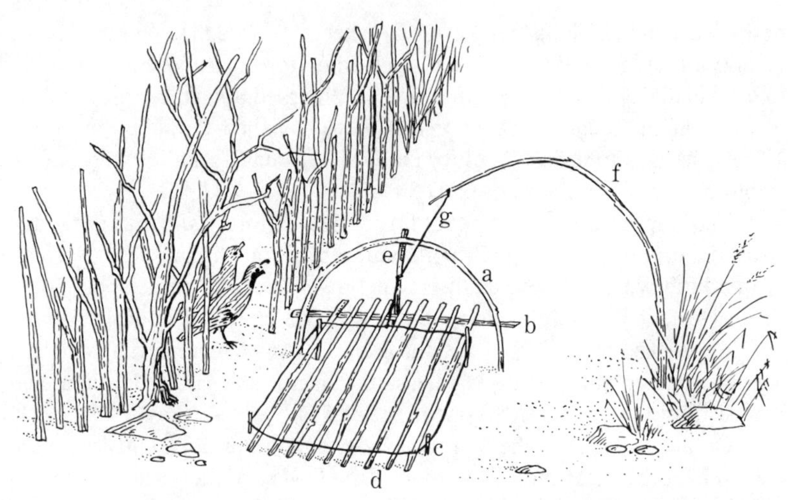

Figure 87. A complicated treadle snare used by the Pomos for capturing quail. (After Barrett, 1952).

anything to pass through this opening, going either way, without stepping on at least one of these treddle bars (d). When the treddle bar went down it depressed the cross-bar (b) which was holding the trigger pin (e). This released the mechanism and allowed the spring pole (f) to operate. The trigger pin (e) was tied firmly onto the string. The pull of the spring pole was directly on this pin. By placing the upper end of the pin against the arch and its lower end against the cross-bar, the tension of the spring pole held the cross-bar against the sides of the arch. The downward pressure on one or more of the treddle bars depressed the cross-bar which thus sank below the lower end of the trigger pin. This tripped the mechanism and allowed the spring pole to fly up, and to take the noose up. The four pegs used to hold the noose open were so slanted that the noose easily slipped off and tightened around the legs of the victim. The chief office of the four pegs was to insure that the noose would not be dislodged by a breeze or by some other accidental means.

It will be remembered that the loop of the snare was spread out flat on the platform made of the several treddle bars; hence, when the bird stepped on any one or more of these treddle bars he was stepping inside the noose. As the spring pole flew up it drew the noose tightly around one or both legs of the bird and hoisted it into the air where it hung fluttering until the arrival of the hunter, who visited his traps usually about mid-forenoon and again after sundown. Informants say that it is the habit of the valley quail to come out of the brush early in the morning and again early in the evening.

The Yuki had a method of hunting quail which is not mentioned elsewhere in the literature. They built a corral four feet in diameter and a foot high, and used grain to lure the birds inside. Once inside the enclosure they were clubbed (Foster 1944:163).

Another technique used in capturing quail appears to have been unique to southern California; it involved the running (wearing) down of quail.

The Cupeno, Luiseno, Diegueno, and Chemehuevi practiced this method of capture. Drucker (1937a:47) notes that: ''Quail were run down in wet, cold weather when their feathers were too wet for them to fly far.'' Sparkman (1908:199), in writing of the Luiseno, states: ''During a prolonged period of cold rainy weather they [quail] become chilled so that they cannot fly far; when in that condition they were formerly run down by boys.'' This method was apparently not used in more northerly areas of the state.

The use of fire at night as an aid in hunting quail is also reported, and it appears to have been used most commonly in the southern area of California among the Serrano, Mountain Cahuilla, Luiseno, Chemehuevi, and Mountain Diegueno. It is unclear if torches were used by the Pomo, but Loeb (1926:165) notes: ''Quail were also hunted at night with oversize blunt arrows having no obsidian point. The men surrounded the quail roost and shot off the quail.'' The southern California method was different, involving the use of clubs. Driver (1937:111) records the use of fire at night for bird hunting, especially quail, among the Monache, Yokuts, and Bankalachi of the southern Sierra Nevada, but no detail is given as to how this was done. For the Tubatulabal, Voegelin (1938:13) describes the method as follows: ''Lighted torches, made from sticks on which pitch had been smeared, [were] waved under trees where quail were roosting at night; as birds flew down they were easily clubbed.'' Sparkman (1908:199) also details this method of capture for the Luiseno: ''The valley quail, found in great numbers in the San Luis Rey valley and adjacent country, even to the summit of Palomar, have always been eaten. They were formerly killed with the bow, and were also hunted at night with fire, dry stems of cholla cactus being set on fire and used to attract them; when they flew towards the light they were knocked down with sticks.'' In other areas, they were shot at as they roosted in trees (Drucker, 1937a:47).

The only other method of capture seems to have been with bow and arrow. We have no ethnographic information on the coast area from San Francisco Bay southward regarding the capture of quail (cf. Harrington 1942:6). However, La Perouse (1798) and Menzies (see Grinnell, 1932) described the use of the bow and arrow in shooting birds along the coastal area. The bow and arrow was also apparently used at night by some groups in the southern Sierra Nevada, as already mentioned. Arrow tips were made of wood rather than stone. Foster (1944:163) describes the Yuki arrows used for quail hunting as follows: ''Special arrows [were used], with a crossed stick point four inches in diameter in place of a flint point. A hunter shooting into a covey could stun three or four birds with one shot, which he picked up before they recovered.'' The Nisenan used a similar type of arrow (Beals 1933:349).

Quail calls were also used (Kroeber, 1961:193), and some groups in the central Sierra used ''songs'' to attract the quail (Aginsky, 1943:396). However, we have no description of the sounds or how they were used.

Belding (1892:121) describes a method which, so far as I know, is peculiar to the Death Valley region:

Lieutenant Birnie, in Geographical Surveys West of the One Hundredth Meridian, describes peculiar blinds that the Indians near Death Valley made, just by springs and artificial ponds, for the purpose of killing quail and other birds. The blinds had the general appearance of beehives; were made of rushes and small boughs interlaced, with an opening for entrance on one side away from the spring. The inside was large enough to seat one person. There was a small hole on the side toward the water through which the arrow was shot. A string was attached to the arrow, and repeated shots could be made with it without alarming the game.

USES OF CALIFORNIA QUAIL IN ABORIGINAL CALIFORNIA

Quail were principally used for food by native Californians. The birds were roasted on coals after capture. In other areas, where large numbers of birds were obtained in drives, the birds were preserved for future use. The Central Miwok in the area of Tuolumne hung quail in trees for about four days or the duration of a quail drive. Quail were dried and stored for winter use among the Miwok (Aginsky, 1943:451), or their flesh was pounded up with salt and thus preserved by the Pomo (Loeb, 1926:165). Quail soup was part of the diet. Quail eggs were also eaten by most groups. These were normally stone-boiled by dropping hot stones into water in baskets with the eggs. The Pomo wrapped eggs in grass and baked them in ashes. Among some groups, such as Yokuts, Serrano, Yuma, Desert Cahuilla, and others, the eggs of quail were taboo to the young. No reason is given for this. Many young Yokut girls would not eat quail for fear that later, when their hair grew long, quail would come and nest in it. The Hupa of northwestern California ate Mountain Quail but did not eat California Quail. The reason given was that the California Quail were believed to spend the day gambling in the underground areas which are the homes of the dead, and their stakes were believed to be the souls of the living (Goddard, 1903:23).

Quail feathers were used by some groups in central California, especially the Pomo, in making delicate feathered baskets (Fig. 88). Quail topknots were used either as a border or scattered around the baskets; they were never used to completely cover the baskets. The plume of the male California Quail is longer and was more valued than that of the female, but both were used (Barrett, 1908:142). Occasionally, the long thin plumes of the Mountain Quail were used in place of the shorter club-shaped plumes of the California Quail. The Yokuts and Miwok also used quail feathers on their baskets at times, but this was rare. George Gibbs journeyed through northern California in 1851, and he described Pomo women in the Clear Lake area wearing hats with woodpecker decoration and edged with tufts of blue quail feathers (Heizer, 1972:9). The Miwok had a quail dance known as the hekeke, but no detail is available.

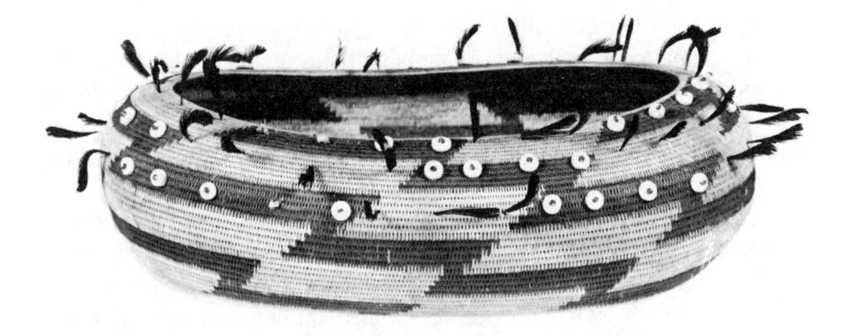

Figure 88. A Pomo basket decorated with top-knots of male California Quail. Lowie Museum of Anthropology photo.

INDIAN NAMES FOR QUAIL

Se-kah-ke, tah-kah-kah, ca-ka-ka, kah-kah-tah, and other variations are some of the terms employed by many separate native Californian groups for the California Quail. There are a number of other words used by other groups, but most used a term resembling se-kah-ke. This is interesting, as the name seems to be onomatopoetic, sounding like the call of the California Quail, which is ca-ca'-cow. There could be little other reason for the widespread distribution of the term other than as an imitation of the quail call.

Linguistic records listing the name for California Quail also give an idea of the range of the bird in native California. However, these cannot be too heavily relied upon, as words can be borrowed and will spread rapidly. More indicative of the actual presence of the California Quail were the special traps and other devices used to capture the birds. For one group, however—the Coast Yurok of the Trinidad Bay area of the northwest coast of California—linguistic notes do contain some information on the distribution of the birds in native California. It was noted that the California Quail had no name as yet because it was a recent arrival to the area and a term had not been given to it. Quail apparently did not occur in the heavily wooded coastal zone until lumbering began. The Yurok subsequently developed a special method of quail capture, however, consisting of a fish net draped over an A-frame.

Words for the California Quail spread far to the east into Nevada and beyond what is thought to be the native distribution of the species. These linguistic records were collected after the turn of the century, however, and by that time the quail had been in western Nevada for forty years. There were no special methods for capturing the quail in this area, so it seems probable that it was recently introduced.

CONCLUSIONS

To many native aborigines, the California Quail was an important part of the diet, supplementing larger mammals, fish, roots, seeds, nuts, and other foods. The birds were so sought after in some areas, especially in the northern half of the state, that special devices were developed solely for the purpose of capturing quail. In the central area of the state, there were professional quail hunters, which emphasizes the importance of the birds to Indians of the area. The construction of long fences up to one-half mile in length also indicates the significance of quail to groups in the northern area of the state. Such devices rewarded native Californians with up to fifty birds a day, which would provide for immediate needs as well as stores for winter.

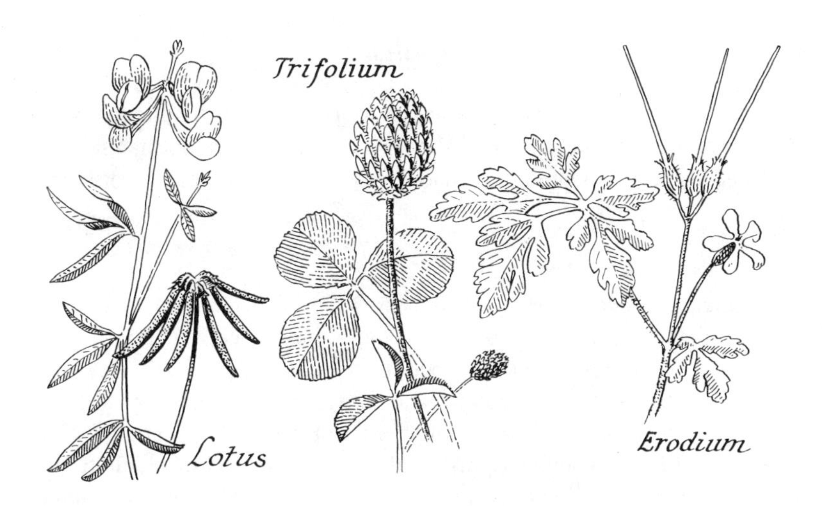

APPENDIX B
FOODS OF THE CALIFORNIA QUAIL

BY BRUCE M. BROWNING
Wildlife Investigations Laboratory
California Department of Fish and Game, Sacramento

Over the years, the Food Habits Section of the Wildlife Investigations Laboratory has analyzed the crop contents of 2,352 California Quail. Six of the samples were collected in the course of intensive studies of the species in various parts of California, and in these instances the data could be subdivided to show changing food habits throughout the year. Tables 18 to 23 are of maximum usefulness in depicting dietary differences between seasons, as well as between areas.

The other 13 samples were derived entirely from birds taken during the legal hunting season, in fall and winter. Most of these samples are small, although a few run over 50 birds. Data in the resulting tables (24 to 36) are only suggestive of local differences in food habits.

The plants enumerated in Tables 18 to 36 are listed only by common or vernacular name. Table 37 gives a complete list of the scientific as well as the vernacular names of all plant species mentioned in the other tables.

Unless otherwise stated, entries in the tables refer to plant seeds or fruits. When greens or leafage are involved, they are so categorized. For example, an entry designated as "Clover" means seeds of plants in the genus *Trifolium*. Greens from the same plants would be labeled "Clover leafage."

Animal foods are listed collectively as "insects," although some arthropods are included (spiders, sow bugs, etc.) that are not members of the Class Insecta. No attempt is made to classify the insects into orders or finer taxa.

SIGNIFICANT FOOD ITEMS

The principal foods eaten by California Quail are supplied by the more common plants comprising the "California Annual Type." This vegetative complex is found throughout the Sacramento and San Joaquin valleys, the Sierra foothills, and in the coastal and inner coastal ranges and valleys. The "California Annual Type," described by Heady (1956), is a principal component of the woodland-grass and woodland-chaparral habitats that make up almost 9 million acres in California (Calif. Dept. Fish and Game, 1965).

This association is dominated by exotic Mediterranean plants such as annual grasses, filarees, and weedy legumes. However, a number of native species useful for quail food persist in the association, particularly legumes. It is the seeds and green leafage of the legumes—the lotuses, clovers, and lupines; filarees, broad-leaf, red-stemmed, and white-stemmed; and annual grasses, bromegrasses, bluegrasses, and fescues—that furnish the bulk of the food eaten by California Quail over most of their range in California (53.2% to 91.3%—70% on the average—from principal samples). Locally and from year to year, other seed items such as turkey mullein, chickweeds, popcorn flowers, thistles, and acorn fragments contribute significantly to the diet over much of the quail range. Waste grain from cultivated barley, wheat, oats, and even rice (Butte County) is significant quail food in some valley and coastal areas. In marginal quail habitats in the Great Basin and in southern California, quail are able to supplement their diet significantly with fruits, buds, and catkins of shrubs such as the sumacs, sagebrush, rabbitbrush, and juniper. Other shrubs whose fruits and seeds contribute to the quail diet include red-berry, manzanita, buck brush, poison oak, and California buckwheat. Plant galls and mistletoe berries are sought by quail at certain times of the year.

California Quail take only small amounts of animal food (1 to 6%), generally in the spring and summer months. Examples are millipedes, mites, spiders, snails, and insects of virtually every order.

Few data are available on the food habits of very young California Quail. Preliminary results from a U.S. Forest Service quail study now being concluded on the San Joaquin Experimental Range in Madera County indicate that young quail may not initially take as high a percentage of animal foods in their diet as other gallinaceous birds. Under the age of 3 weeks, however, California Quail chicks do use significant amounts of animal food, but the intake decreases until, at the age of 13–16 weeks, their diet is very similar to that of the adults.

SEASONAL FOOD HABITS

The diet of the California Quail changes dramatically with the seasons. Data from the 6 principal sampling areas (Tables 18 to 23) are presented seasonally, as follows: spring (March, April, and May); summer (June, July, and August); fall (September, October, and November); and winter (December, January, and February).

Analysis of the principal sampling areas indicates that in the spring the California Quail consumes a high percentage (about 35%) of green leafage, mostly the tender leaves of the clovers, filarees, and grasses. The seeds of the same plants furnish another third of the diet. The balance is comprised of the seeds of spring annuals such as chickweed, miner's lettuce, red maids, geranium, buttercup, buckthorn weed, and early popcorn flowers. Acorn fragments are sometimes used in the spring.

In the summer, legume seeds together with the seeds of filaree and annual grasses make up over two-thirds of the diet. Turkey mullein, popcorn flower, buckthorn weed, windmill pink, thistles, tarweeds, and California poppy are summer annuals that constitute the balance of the summer diet over much of the California Quail's range.

Food habits in the fall are similar to those in the summer, with legumes, filaree, annual grass seeds, and/or cultivated grains supplying the staple diet (over 50%). In areas or years when acorns are available, fragments of these large seeds are sought out by quail. Use of turkey mullein generally peaks in the fall, and other late summer annuals such as vinegar weed, gambleweed, tarweed, and thistles supplement the diet during the fall months.

The ranges vegetated with the "California Annual Type" green-up in response to late fall or winter rains (it takes approximately ½ inch of rain to induce germination). California Quail respond to this event immediately by eating green leafage. In the crops examined, green leafage was principally that of the clovers, filarees, and annual grasses, in that order of preference or selectivity. In the principal sampling areas, 43.5% of the winter food was green leafage; only a little over 1% of that was grass leafage. Seeds of legumes, filaree, cultivated grains, and/or annual grass seed still constituted a third of the diet, substantiating the designation of these as "staple" quail foods. Green leafage is absent from some of the winter samples: namely, Inyo County (1939–40), western Kern and Santa Barbara Counties (1948), and Shasta County (1959); and is low in others: Monterey County (1960), Nevada County (1938), and Butte County (1961). However, in valley areas and foothills of the Sierra Nevada (Madera County), one can tell exactly when the range "greened up" by noting the first date that green leafage appears in the quail crops examined. However, as noted in Chapter 9, the palatability and nutritive value of greens may differ from year to year, depending on growing conditions and the vigor of the plants. Green foods derived from rapidly growing annuals appear to contribute to the reproductive energy of quail, whereas stunted plants of the same species may contain isoflavones (and perhaps other compounds) that inhibit reproduction. Knowledge of the significance of green foods in the quail diet is still inadequate.

When available, acorns are important winter food for the California Quail. Results from the 10-year, cooperative quail study now being concluded at the San Joaquin Experimental Range (Madera County) have shown that, in a good acorn year, acorn fragments sustain quail not only through the fall and winter, but also into spring as well. Note the high acorn percentages in the Madera County sample, Table 18.

MANAGEMENT IMPLICATIONS OF FOOD HABITS DATA

The principal food items of California Quail are derived from some of the most abundant and ubiquitous plants of the arid California rangelands. In a year of adequate rainfall, these annual forbs are capable of producing enough food to sustain a high quail population. Likewise, in wet years the green leaves of the same plants are consumed avidly by the birds and seemingly contribute to their health and reproductive drive. Conversely, in dry years forb growth is stunted, and neither the greens taken in winter and spring nor the seeds subsequently produced are conducive to high quail populations. Variable rainfall, therefore, is a major factor affecting the availability and nutritive value of quail foods in arid portions of the quail range.

A second factor affecting the availability of quail food is grazing of the foothill ranges by domestic livestock. The forbs that feed quail likewise are among the favorite foods of sheep and cattle, and when grazing is too intensive even the commonest forbs can be consumed to the point where the seed crop is substantially reduced. The impact of grazing is most adverse on arid ranges and particularly in dry years. Conversely, grazing on more humid ranges may open up dense stands of annual grass, favoring the growth of forbs and the general welfare of quail.

In terms of management, there is little that can be done to influence rainfall, but grazing is subject to local manipulation to create a favorable environment for forb growth and seed maturation. Controlled livestock grazing is suggested as a major management tool for improving the food supply of the California Quail.

TABLE 18.
Food habits of 1,229 California Quail collected in Madera County, 1960–72

Spring (123)	*Vol %*	*Fr.*	*Summer (125)*	*Vol %*	*Fr.*	*Fall (426)*	*Vol %*	*Fr.*	*Winter (565)*	*Vol %*	*Fr.*
Clover leafage	27.6	93	Spanish clover	27.9	133	Oak acorns	35.1	289	Oak acorns	34.2	327
Ground lupine	13.8	41	Ground lupine	16.9	111	Turkey mullein	13.2	242	Clover leafage	22.9	417
Common chickweed	7.3	55	Clover seed	8.8	106	Spanish clover	10.3	278	Strigose lotus	10.4	235
Bluegrass	6.5	35	Broad-leaf filaree	5.3	89	Ground lupine	8.9	245	Clover seed	8.3	288
Oak acorns	5.7	21	Turkey mullein	4.7	42	Clover seed	6.1	300	Ground lupine	5.7	217
Sunflower, flor. & seed	4.9	64	Popcorn flowers	4.7	78	Strigose lotus	5.7	214	Filaree leafage	5.1	244
Clover seed	4.1	59	Oak acorns	4.1	15	Gall fragments	3.6	110	Forb leafage	4.0	202
Mouse-ear chickweed	3.8	36	Bluegrass	2.9	16	Clover leafage	3.0	102	Spanish clover	3.0	177
Popcorn flowers	3.3	43	Buckthorn weed	2.8	40	Broad-leaf filaree	3.0	237	Grass leafage	2.2	331
Buckthorn weed	2.8	26	Manzanita	2.6	16	Lotus	2.7	90	Miscellaneous food items	4.2	—
Forb leafage	2.7	60	Red-berry	2.3	14	Forb leafage & seed	2.5	154			
Shining chickweed	2.7	21	Red-stem filaree	1.9	21	Grass leafage	1.4	179			
Smooth cat's-ear	2.6	20	Windmill pink	1.5	73	Miscellaneous food items	4.5	—			
Broad-leaf filaree	2.3	47	Galls	1.3	12						

TABLE 18. Continued

Food habits of 1,229 California Quail collected in Madera County, 1960–72

Spring (123)	*Vol %*	*Fr.*	*Summer (125)*	*Vol %*	*Fr.*	*Fall (426)*	*Vol %*	*Fr.*	*Winter (565)*	*Vol %*	*Fr.*
Miner's lettuce	2.0	51	Miner's lettuce	1.2	25						
Spanish clover	1.3	24	White-stem filaree	1.2	17						
Miscellaneous food items	6.6	—	Clover leafage	1.1	20						
			Strigose lotus	1.1	70						
			Insect fragments	1.0	40						
			Miscellaneous food items	6.7	—						

Vol % = percentage volume
Fr. = Frequency of occurrence

TABLE 19.
Food habits of 114 California Quail collected in Madera County, 1937 (Glading, Biswell, and Smith, 1940)

Spring (29)	*Vol %**	*Summer (33)*	*Vol %*	*Fall (27)*	*Vol %*	*Winter (25)*	*Vol %*
Filaree	16.5	Spanish clover	25.1	Turkey mullein	28.8	Forb leafage	24.7
Forb leafage	15.7	Filaree	22.8	Filaree	17.8	Grass leafage	14.7
Clover leafage	14.0	Turkey mullein	19.1	Clover	17.0	Filaree leafage	11.6
Bluegrass	7.3	Clover	10.3	Spanish clover	15.6	Clover leafage	8.7
Popcorn flowers	7.0	Popcorn flowers	4.8	Forb leafage	9.9	Lily family leafage	8.2
Gilia	6.6	Buckthorn weed	2.7	Tarweed	5.3	Filaree	6.8
Filaree leafage	5.3	Tarweed	2.4	Oak acorns	2.7	Oak acorns	6.3
Clover	4.8	Ground lupine	1.6	Ground lupine	1.1	Turkey mullein	4.3
Grass leafage	3.6	Miscellaneous food items	11.2	Miscellaneous food items	1.8	Spanish clover	4.2
Linanthus leafage	2.9					Gilia leafage	3.8
Spanish clover	2.3					Linanthus leafage	2.9
Lily family leafage	2.2					Clover	2.0
Buckthorn weed	1.9					Popcorn flower leafage	1.0
Animal food items	1.6					Miscellaneous food items	0.8
Buckthorn weed leafage	1.0						
Miscellaneous food items	7.3						

*Frequencies not available.

TABLE 20.
Food habits of 158 California Quail collected in Lake and Mendocino Counties, 1950–51, 1962–65

Spring (28)	*Vol %*	*Fr.*	*Summer (17)*	*Vol %*	*Fr.*	*Fall (66)*	*Vol %*	*Fr.*	*Winter (47)*	*Vol %*	*Fr.*
Clover leafage	33.6	24	Lotus	23.0	13	Oak acorns	18.6	25	Clover leafage	45.0	38
Geranium	21.9	18	Clover	12.0	13	Clover	18.5	49	Filaree leafage	13.1	22
Chickweed	20.3	16	Ryegrass	10.0	5	Turkey mullein	13.4	33	Forb leafage	12.1	36
Lotus	4.1	7	Geranium	8.3	8	Clover leafage	8.1	41	Cult. barley	7.8	5
Oak acorns	3.6	2	Lupine	7.8	5	Vetch	5.6	5	Oak acorns	5.6	13
Clover	2.4	8	Manzanita	6.8	2	Lotus	3.7	15	Bur clover	3.1	9
Lupine	2.3	4	Wild oats	6.2	9	Filaree leafage	3.3	23	Clover	1.9	24
Filaree leafage	2.1	3	Buckbrush	3.4	1	Bur chervil	2.8	20	Turkey mullein	1.6	8
Red maids	1.8	1	Spanish clover	3.4	5	Geranium	2.4	12	Bluegrass	1.4	1
Sunflower stems, leafage	1.8	2	Clover leafage	2.6	3	Forb leafage	2.4	1	Ceanothus	1.4	1
Miner's lettuce	1.7	3	Bluegrass	1.9	4	Cult. barley	2.3	2	Spanish clover	1.2	3
Forb leafage	1.3	13	Insect fragments	1.8	6	Ground lupine	2.1	22	Ground lupine	1.1	4
Red-stem filaree	1.0	9	Popcorn flowers	1.7	1	Buckbrush	2.0	3	Grass leafage	1.1	29
Miscellaneous food items	2.1	—	Ceanothus	1.4	3	Grass leafage	1.9	23	Lotus	1.0	11
			Oak acorns	1.3	1	Bluegrass	1.2	6	Miscellaneous food items	2.6	—
			Forb leafage	1.3	4	Galls	1.2	10			
			Vetch	1.2	2	Red-stem filaree	1.1	15			
			Red-stem filaree	1.2	12	Vinegar weed	1.1	2			
			Soft chess	1.2	6	Bur clover	1.0	15			

	Miscellaneous food items	3.5	—	Wild oats	1.0	4
				Unidentified fruit fragments	1.0	9
				White-stem filaree	1.0	9
				Miscellaneous food items	4.3	—

TABLE 21.

Food habits of 145 California Quail collected in San Luis Obispo and San Benito Counties, 1941, 1949–51, 1966

Summer (12)	*Vol %*	*Fr.*	*Fall (82)*	*Vol %*	*Fr.*	*Winter (51)*	*Vol %*	*Fr.*
Red-stem filaree	41.9	12	Oak acorns	15.5	20	Red-stem filaree	19.0	35
Lupine	15.7	9	Red-stem filaree	15.0	55	Turkey mullein	15.6	19
Canaigre	7.7	2	Ground lupine	12.6	23	Forb leafage	11.0	23
Lily corns	6.7	1	Lupine	7.5	21	Hill lotus	10.5	20
Wild barley	4.5	2	Cult. barley	7.4	14	Cult. barley	6.9	14
Prickly lettuce (trap bait)	4.2	5	Turkey mullein	5.2	33	Vinegar weed	5.8	10
Hill lotus	3.9	7	Popcorn flowers	5.2	23	Clover	5.4	23
White plectritis	3.4	4	Hill lotus	3.7	36	Filaree leafage	5.1	2
Grasshopper fragments	3.3	1	Cult. wheat	3.5	7	California croton	3.8	10
Ants	2.9	6	Forb leafage	2.4	29	Toyon	3.8	4
Red-stem filaree leafage	2.1	1	Vinegar weed	2.4	9	Grass leafage	3.4	32
Melic grass	1.3	2	Buckthorn weed	2.2	21	Lupine	1.9	21
Lotus	1.0	6	Wild oat	1.9	13	Bur clover	1.6	20
Miscellaneous food items	1.4	—	Bassia	1.7	2	Common Lomatium	1.5	14
			Grass leafage	1.6	37	Miscellaneous food items	4.7	—
			Fescue	1.4	16			
			Slender fescue	1.0	22			
			Clover	1.0	31			
			Miscellaneous food items	8.8	—			

TABLE 22.
Food habits of 127 California Quail collected in San Luis Obispo County, 1971–73
(Erwin, Appendix C)

Spring (41)	*Vol %*	*Fr.*	*Summer (57)*	*Vol %*	*Fr.*	*Fall (7)*	*Vol %*	*Fr.*	*Winter (22)*	*Vol %*	*Fr.*
Red-stemmed filaree	17.2	34	Ground lupine	21.1	41	Red-stem filaree	27.0	6	Filaree leafage	22.7	16
Ground lupine	13.8	20	Red-stem filaree	20.6	54	Spanish clover	15.3	4	Forb leafage	13.4	15
Hill lotus	7.2	21	Cult. barley	10.6	13	Hill lotus	13.6	4	Grass leafage	13.4	14
Tessellate fiddle-neck	7.1	17	Cult. wheat	7.7	8	Turkey mullein	12.7	7	Hill lotus	7.1	17
Insect fragments	6.9	34	Miner's lettuce	5.8	16	Ground lupine	10.6	7	Napa thistle	5.7	9
Forb leafage & buds	6.4	23	Hill lotus	3.7	36	Vinegar weed	7.8	2	Oak acorns	4.6	3
Clover leafage	6.3	9	Insect fragments	3.0	27	Tessellate fiddle-neck	5.0	3	Red-flower lupine	4.5	3
Lupine seeds and pods	5.7	17	Gilia	3.0	5	Cult. barley	4.7	1	Cult. barley	4.4	5
Filaree leafage	3.5	11	Forb leafage & buds	2.6	34	Common fiddle-neck	2.0	4	Turkey mullein	2.4	3
Common fiddle-neck	2.6	17	Napa thistle	1.8	24	Miscellaneous food items	1.3	—	Red-stem filaree	2.3	12
Miner's lettuce leafage	2.4	4	Clover	1.6	29				Deerweed	2.1	1
Miner's lettuce	2.2	6	Unidentified sun-flower	1.6	7				Ground lupine	2.0	3
Honeysuckle	2.0	5	Lupine	1.5	12				Clover	1.8	9

TABLE 22. Continued

Food habits of 127 California Quail collected in San Luis Obispo County, 1971–73

(Erwin, Appendix C)

Spring (41) *Fall (7)*		*Vol %* *Vol %*		*Fr.* *Fr.*		*Summer (57)* *Winter (22)*	*Vol %* *Vol %*		*Fr.* *Fr.*	
Shiny chickweed	1.9	1	Rusty popcorn flowers	1.3	20			Vinegar weed	1.8	8
Lotus seeds & pods	1.7	9	Cult. oats	1.2	3			Bur clover	1.8	2
Cult. barley	1.6	4	Common fiddle-neck	1.1	18			Lotus leafage	1.7	3
Grass leafage	1.3	8	Turkey mullein	1.0	13			Cult. wheat	1.7	2
Carrot family leafage	1.1	1	Miscellaneous food items	10.8	—			Lupine leafage	1.6	1
Gooseberry or currant	1.0	1						Loco weed	1.4	4
Wild barley	1.0	2						Angelica	1.3	3
Miscellaneous food items	7.1	—						Miscellaneous food items	2.3	—

TABLE 23.
Food habits of 102 California Quail collected in Santa Cruz County, 1935
(Summer, 1935)

Spring (15)	*Vol %**	*Summer (31)*	*Vol %*	*Fall (32)*	*Vol %*	*Winter (24)*	*Vol %*
Forb leafage	44.0	Cult. barley	17.7	Bur clover	37.2	Bur clover	22.4
Bur clover	17.8	Italian ryegrass	13.1	Forb leafage	9.9	Forb leafage	16.0
Red-stem filaree	9.1	Lupines	9.3	Cult. barley	6.0	Lupines	14.6
Acacia	7.7	Tarweed (*Madia*)	6.9	Poison oak	6.0	Cult. barley	13.4
Clover	5.9	Red-stem filaree	6.8	Red-stem filaree	5.9	Red-stem filaree	7.3
Buttercup	3.5	Forb leafage	5.6	Soft chess	5.7	Soft chess	5.7
Italian ryegrass	3.4	Napa thistle	5.3	Lupines	4.4	Buttercup	5.3
Soft chess	3.4	Tarweed (*Hemizonia*)	3.1	Tarweed (*Hemizonia*)	4.3	Cult. oats	3.9
Chickweed	3.3	Spanish clover	2.7	Acacia	3.1	Acacia	3.1
Insect fragments	1.1	Acacia	2.5	Napa thistle	2.1	Mustard	1.9
Miscellaneous food items	0.8	Scarlet pimpernel	2.3	Italian ryegrass	1.7	Italian ryegrass	1.3
		Thimbleberry	2.2	Gambleweed	1.6	Gambleweed	1.3
		Soft chess	2.0	Wild geranium	1.4	Toyon	1.1
		Wild geranium	2.0	Scarlet pimpernel	1.3	Miscellaneous food items	2.7
		Lotus	1.7	Snowberry	1.3		
		Galls	1.4	Lotus	1.1		
		California poppy	1.2	Sheep sorrel	1.1		
		Elderberry	1.1	Toyon	1.0		
		Bur clover	1.0	Miscellaneous food items	4.9		
		Miner's lettuce	1.0				
		Insect fragments	1.0				
		Miscellaneous food items	10.1				

*Frequencies not available.

TABLE 24.
Food habits of 51 California Quail collected in Lassen, Modoc, and Siskiyou Counties, 1948–49, 1959

Winter (51)	*Vol %*	*Fr.*
Grass leafage	25.7	44
Cheatgrass	20.5	50
Juniper catkins	18.1	23
Tumbling mustard	9.9	34
Lupine	6.7	12
Filaree leafage	6.1	17
Woodland Star winter buds	2.0	8
Lotus	2.0	6
Yellow pine nuts	1.4	2
Buckwheat	1.3	8
Black medick	1.2	8
Red-stem filaree	1.2	17
Wild barley	1.0	1
Miscellaneous food items	2.9	—

TABLE 25.
Food habits of 10 California Quail collected in Shasta County, 1959

Winter (10)	*Vol %*	*Fr.*
Mistletoe berries	30.4	4
Turkey mullein	19.7	3
Red-stem filaree	17.5	4
Cultivated wheat	13.9	2
Oak acorns	8.1	4
Cultivated barley	5.7	2
Rip-gut	4.5	8
Miscellaneous food items	0.2	—

TABLE 26.
Food habits of 14 California Quail collected in Nevada County, 1938

Winter (14)	*Vol %*	*Fr.*
Lotus	55.0	13
Spanish clover	19.8	5
Oak acorns	17.7	9
Ground lupine	2.8	6
Forb leafage	2.2	8
Bur clover	1.3	4
Grass leafage	1.1	8
Miscellaneous food items	0.1	—

TABLE 27.
Food habits of 41 California Quail collected in Yuba County, 1972–73

Fall–Winter (41)	Vol %	Fr.
Bur clover	36.9	39
Clover leafage	22.5	38
Oak acorns	22.2	25
Filaree leafage	7.6	38
Lupine	3.5	15
Clover	2.1	13
Mistletoe fruit	1.7	9
Poison oak	1.3	5
Other food items	2.0	—

TABLE 28.
Food habits of 24 California Quail collected in Butte County, 1938, 1961

Fall, 1938 (18)	Vol %	Fr.	Winter, 1961 (6)	Vol %	Fr.
Yellow-star thistle	32.0	13	Vetch	30.0	5
Milk thistle	18.1	13	Italian rye grass	19.8	6
Gamble weed	11.3	13	Broad-leaved filaree	15.8	6
Cultivated rice	11.1	2	Yellow-star thistle	14.8	6
Oak acorn	7.1	12	Wire grass	14.3	6
Lotus	4.9	2	Common chickweed	1.7	5
Forb leafage	3.9	7	Clover leafage	1.7	4
Bur clover	3.8	18	Bur clover	1.2	5
Wild grape	3.0	4	Other food items	0.7	—
Grass leafage	2.7	5			
Snake-root	1.4	13			
Other food items	0.7	—			

TABLE 29.
Food habits of 9 California Quail collected in Marin County, 1950

Winter (9)	Vol %	Fr.
Clover leafage	53.3	6
Forb leafage	23.8	5
Yellow sweet clover	15.6	3
Grass leafage	3.3	2
White sweet clover	1.7	1
Sedge	1.2	4
Lupines	1.1	2

TABLE 30.
Food habits of 69 California Quail collected in Monterey County, 1960

Winter (69)	*Vol %*	*Fr.*
Red-stem filaree	36.5	65
Oak acorns	10.0	19
White sage	7.4	25
Scarlet pimpernel	6.7	26
Napa thistle	6.6	50
Strigose lotus	5.6	20
Wild lilac	4.1	20
Manzanita berries	3.6	13
Lupines	2.5	19
Forb leafage	2.4	18
Vetch	2.1	2
Deerweed	1.3	12
Clovers	1.3	18
Galls	1.0	9
Ryegrass	1.0	6
Miscellaneous food items	7.9	—

TABLE 31.
Food habits of 10 California Quail collected in Fresno County, 1971–72

Winter (10)	*Vol %*	*Fr.*
Red-stem filaree	32.6	10
Forb leafage	22.6	10
Wild carrot	17.5	8
Grass leafage	15.3	9
Turkey mullein	8.9	9
Hill lotus	1.5	8
Other food items	1.6	—

TABLE 32.
Food habits of 56 California Quail collected in Inyo County, 1939–40

Winter (56)	*Vol %*	*Fr.*
Russian thistle	25.7	33
Sagebrush	19.6	33
Rabbitbrush	16.2	38
Locust	12.1	16
Maize	9.6	10
White sweet clover	6.5	19
Sunflower	2.3	4
Cultivated wheat	1.8	1
Rough pigweed	1.6	7
Spider plant	1.0	3
Miscellaneous food items	3.6	—

TABLE 33.
Food habits of 14 California Quail collected in Kern and Santa Barbara Counties, 1948

Winter (14)	*Vol %*	*Fr.*
Turkey mullein	43.1	9
Hill lotus	11.9	4
Clover	10.1	11
Fescue	8.5	9
Stephanomeria	4.6	1
Rough pigweed	4.3	1
Mallow	3.6	2
Buckthorn weed	2.7	3
Telegraph weed	2.1	2
Jackass clover	2.0	1
Vegetative fragments	1.6	2
Other food items	5.5	—

TABLE 34.
Food habits of 9 California Quail collected in San Bernardino County, 1966

Winter (9)	*Vol %*	*Fr.*
Filaree leafage	48.4	8
Oak acorns	8.9	2
Grass leafage	7.4	8
California buckwheat	6.9	2
Forb leafage	6.7	5
Red-stem filaree	6.4	4
Phacelia	5.6	2
Lupine	5.1	2
Tumbling mustard	3.9	2
Miscellaneous food items	0.7	—

TABLE 35.
Food habits of 145 California Quail collected on Catalina Island, Los Angeles County, 1949

Winter (145)	*Vol %*	*Fr.*
Cultivated barley	19.4	49
Red-stem filaree	16.2	123
White-stem filaree	12.8	113
Wild oats	9.1	97
Wild barley	8.1	102
Bur clover	6.7	126
Laurel-sumac	5.1	54
Bromegrass	4.3	110
Snake-root	4.1	93
Grass leafage	3.9	129
Lupines	2.6	57
Australian saltbush	2.3	62
Lotus	1.1	57
Miscellaneous food items	4.3	—

TABLE 36.
Food habits of 25 California Quail collected in San Diego County, 1938

Winter (25)	*Vol %*	*Fr.*
Forb leafage	27.3	10
Lemonade-berry fruits	21.9	6
Laurel sumac fruits	17.2	7
Bur clover	12.6	8
Filaree leafage	10.7	21
Turkey mullein	3.2	10
Moss leafage	2.5	8
Lupine	1.6	6
Brome grass	1.0	10
Miscellaneous food items	2.0	—

TABLE 37.
Vernacular and scientific names of plants eaten by California Quail

Acacia	*Acacia* sp.
Acorns	*Quercus* sp.
Barley, cultivated	*Hordeum vulgare*
Barley, wild	*Hordeum* sp.
Bluegrass	*Poa annua*
Brome grass	*Bromus* sp.
Buckbrush	*Ceanothus cuneatus*
Buckthorn weed	*Amsinckia douglasiana*
Buckwheat, California	*Eriogonum fasciculatum*
Buckwheat, wild	*Eriogonum* sp.
Buttercup	*Ranunculus* sp.
Canaigre	*Rumex hymenosepalus*
Carrot, wild	*Daucus* sp.
Cat's-ear, smooth	*Hypochoeris glabra*
Cheat grass	*Bromus tectorum*
Chervil, bur	*Anthriscus scandicina*
Chickweed, common	*Stellaria media*
Chickweed, mouse-ear	*Cerastium viscosum*
Chickweed, shining	*Stellaria nitens*
Clover	*Trifolium* sp.
Clover, bur	*Medicago hispida*
Clover, jackass	*Wislizenia refracta*
Clover, Spanish	*Lotus purshiana*
Clover, white sweet	*Melilotus alba*
Clover, yellow sweet	*Melilotus indica*
Croton, California	*Croton californicus*
Deer-weed	*Lotus scoparius*
Dock	*Rumex* sp.
Elderberry	*Sambucus* sp.
Fescue	*Festuca* sp.
Filaree	*Erodium* sp.
Fiddleneck	*Amsinckia* sp.
Filaree, broad-leaved	*Erodium botrys*
Filaree, red-stem	*Erodium cicutarium*
Filaree, white-stem	*Erodium moschatum*
Gambleweed	*Sanicula menziesii*
Geranium, wild	*Geranium disectum*
Gilia, straggling	*Gilia gilioides*
Gooseberry	*Ribes* sp.
Grape	*Vitis* sp.
Grass family	Gramineae
Hog-fennel	*Lomatium* sp.
Honeysuckle	*Lonicera* sp.
Juniper	*Juniperus* sp.
Lemonade-berry	*Rhus integrifolia*

TABLE 37. Continued
Vernacular and scientific names of plants eaten by California Quail

Lettuce, miner's	*Montia perfoliata*
Lettuce, prickly	*Lactuca scariola*
Lily family	Liliaceae
Linanthus	*Linanthus* sp.
Locoweed	*Astragalus* sp.
Locust	*Robinia pseudoacacia*
Lotus	*Lotus* sp.
Lotus, hill	*Lotus humistratus* or *subpinnatus*
Lotus, strigose	*Lotus strigosus*
Lupine	*Lupinus* sp.
Lupine, ground	*Lupinus bicolor*
Maize	*Zea mays*
Mallow	*Malva* sp.
Manzanita	*Arctostaphylos* sp.
Medic	*Medicago* sp.
Medic, black	*Medicago lupulina*
Melic grass	*Melica* sp.
Mistletoe	*Phoradendron* sp.
Moss	Lycopodinae
Mullein, turkey	*Eremocarpus setigerus*
Mustard family	Cruciferae
Mustard, tumbling	*Sisymbrium altissimum*
Oats, cultivated	*Avena sativa*
Oats, wild	*Avena fatua*
Plectritis	*Plectritis* sp.
Phacelia	*Phacelia* sp.
Pigweed, rough	*Amaranthus retroflexus*
Pimpernel, scarlet	*Anagallis arvensis*
Pine, yellow	*Pinus ponderosa*
Pink, windmill	*Silene gallica*
Poison oak	*Rhus diversiloba*
Popcorn flower	*Plagiobothrys* sp.
Poppy, California	*Eschscholzia californica*
Rabbitbrush	*Chrysothamnus* sp.
Red-berry	*Rhamnus crocea*
Red maids	*Calandrinia caulescens*
Rice, cultivated	*Oryza sativa*
Rip-gut	*Bromus rigidus*
Rye grass	*Lolium* sp.
Rye grass, Italian	*Lolium multiflorum*
Sage, white	*Salvia apiana*
Sagebrush	*Artemisia tridentata*
Saltbush, Australian	*Atriplex semibaccata*
Sedge family	Cyperaceae
Snake-root	*Sanicula* sp.

TABLE 37. Continued
Vernacular and scientific names of plants eaten by California Quail

Snowberry	*Symphoricarpos* sp.
Soft chess	*Bromus mollis*
Sorrel, sheep	*Rumex acetosella*
Spider plant	*Cleome lutea*
Stephanomeria	*Stephanomeria* sp.
Sumac, laurel	*Rhus laurina*
Sunflower family	Compositae
Tarweed	*Madia* sp., *Hemizonia* sp.
Telegraph weed	*Heterotheca grandiflora*
Thimble-berry	*Rubus parviflorus*
Thistle, milk	*Silybum marianum*
Thistle, Napa	*Centaurea melitensis*
Thistle, Russian	*Salsola kali*
Thistle, yellow star	*Centaurea solstitialis*
Toyon	*Photinia arbutifolia*
Vetch	*Vicia* sp.
Vinegar weed	*Trichostema lanceolatum*
Wheat, cultivated	*Triticum aestivum*
Wild lilac	*Ceanothus* sp.
Woodland star	*Lithophragma affinis*

APPENDIX C
EFFECTS OF DIFFERING RAINFALL ON BREEDING OF CALIFORNIA QUAIL IN AN ARID ENVIRONMENT

BY MICHAEL J. ERWIN
Department of Forestry and Conservation
University of California, Berkeley

It has long been recognized that the California Quail in the more arid parts of its range reproduces prolifically in wet years and is comparatively unproductive in dry years. From a teleologic standpoint, the logic of this arrangement is clear enough. Rain induces a robust growth of forbs whose seeds serve to nourish growing young and to support an enlarged population over winter. Conversely, drought leads to a paucity of forb growth with resultant low food resources even for the adult quail, let alone for a big crop of young. It is obviously advantageous for quail to expend the enormous physiologic energy of reproduction only in years when the prospective food supply will support the resultant progeny. In dry years, the best strategy is to conserve resources and to await more favorable circumstances for breeding. This is precisely what quail do. But the biological mechanism which regulates variable reproductive behavior is not understood.

I was given the opportunity to study this phenomenon in connection with the Quail Project of the California Academy of Sciences, financed by a grant from Peter McBean. The site selected for the field study was the vicinity of Shandon, San Luis Obispo County, where a great backlog of information on quail ecology had been accumulated by Ian McMillan, whose ranch is situated a few miles southeast of Shandon. McMillan annually recorded the sex and age composition of quail populations since 1949 (see Table 8), and he published some of his observations on the relationship of rainfall to forb growth and to quail reproduction (McMillan, 1964). Subsequently, Francis (1970) analyzed in more detail the weather records from the Shandon area and related these to quail reproductive success. The purpose of my project was to follow in detail the physiologic and behavioral aspects of quail reproduction through a period of time which hopefully would include dry and wet years.

The field study was initiated in 1971 and extended through 1973. By great good fortune, 1972 proved to be a dry year with poor quail reproduction and 1973 was a wet year with prodigious reproduction. The report which follows compares the social and reproductive behavior of the birds during these two contrasting seasons, leading to some suggested ideas on how rainfall and plant growth may regulate quail reproduction.

STUDY AREA

The study area is situated in northeastern San Luis Obispo County, south of the town of Shandon. The Ian McMillan Ranch, nine road miles southeast of Shandon, marks the northernmost portion of the study area and was the locality that served as the base from which the research was conducted. Within the study area, in addition to the McMillan Ranch, are six other sub-units (named in accordance with local tradition) where the majority of observations were made and birds were collected. These are Iver's Windmill, the Deerfield, the Horsepasture, the Pines, the Fernandez, and—within the National Forest—the Fernandez Road plot. Figure 89 depicts the location of these localities.

The soils of the northern half of the study area are rich clay loams derived from the Paso Robles formation. This area is dry-farmed for wheat or barley on the flats and grazed on the slopes. The floors of the San Juan and Camatta drainages are irrigated and support fields of alfalfa, sugar beets, and grape vineyards. Woody vegetation is sparse.

South of the Pines, the soils are more sandy and less fertile. There are extensive areas of oak woodland and chaparral, all of which are grazed.

McMillan (1964) has described the topography and climate. Weather data were provided by the San Luis Obispo County Engineer's Department as recorded on the Highland Ranch (climatological station 122) located at the eastern boundary of the McMillan Ranch. The cumulative annual rainfall for 1972, 1973, and for a 13-year mean are presented in Fig. 90.

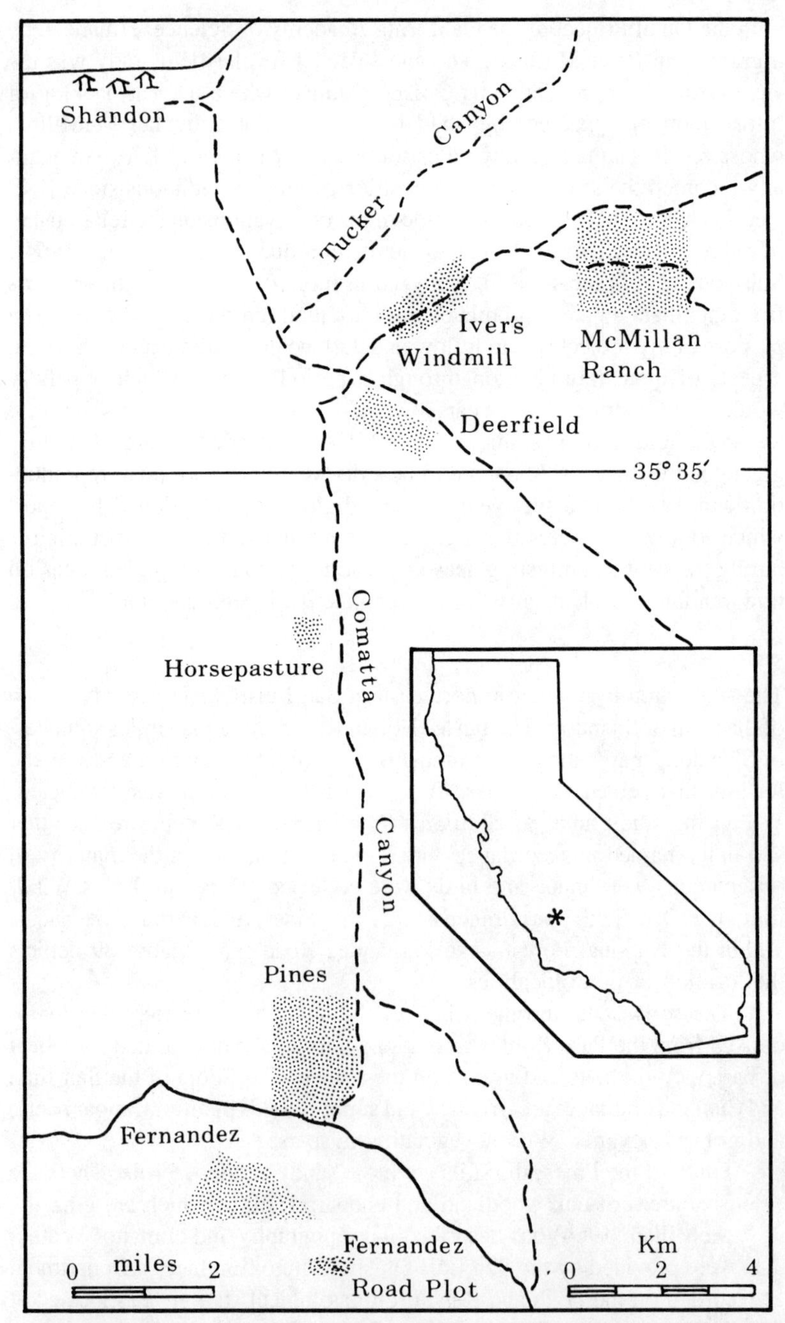

Figure 89. Map of the study area south of Shandon. Stippled areas depict principal study sites.

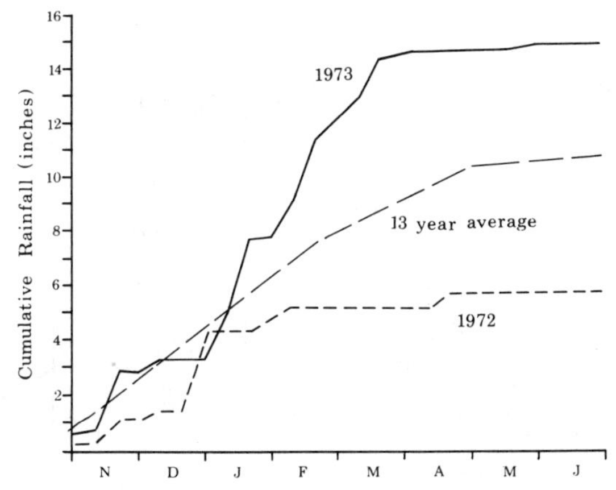

Figure 90. Cumulative rainfall records at Shandon for 1972 and 1973 in comparison with average rainfall.

STUDY METHODS

To obtain quail crops and gonads for analysis, I shot birds during a two-hour period in mid-morning or late afternoon, after the morning and evening feeding periods, when their crops were full. Each month, from December 1971 through August 1972 and from December 1972 through October 1973, I collected five quail of each sex. When pairing of the birds became evident, every effort was made to collect members of pairs. Adults with young were not collected.

Within two hours after shooting, each bird was weighed, dissected, and the reproductive organs measured, weighed, and preserved. Half of the crops were preserved in formalin for quantitative food analysis, and the others were frozen for chemical analysis. One wing of each bird was saved for later study of molt.

Observations of social and reproductive behavior were made during my travels over the study area, but concentrated observation was limited largely to McMillan's "home covey," using my car as a blind.

Each year I recorded the general range condition by noting timing of seed germination, flowering, seed set, and seed shatter of key annual plant species. Key species were chosen on the basis of their importance as quail food in green and/or seed-producing stages.

TESTICULAR CYCLES—1972 AND 1973 COMPARED

Figure 91 depicts the recrudescence and regression of testis size in quail of the Shandon area. Lewin (1963) noted that testes larger than 7.5 to 8.0 mm. produced viable sperm, so I am assuming that the effective breeding period for each year could be determined for testis size. As Figure 91 shows, the testes of Shandon quail began developing during both years in early February and reached breeding size by late February or early March—actually

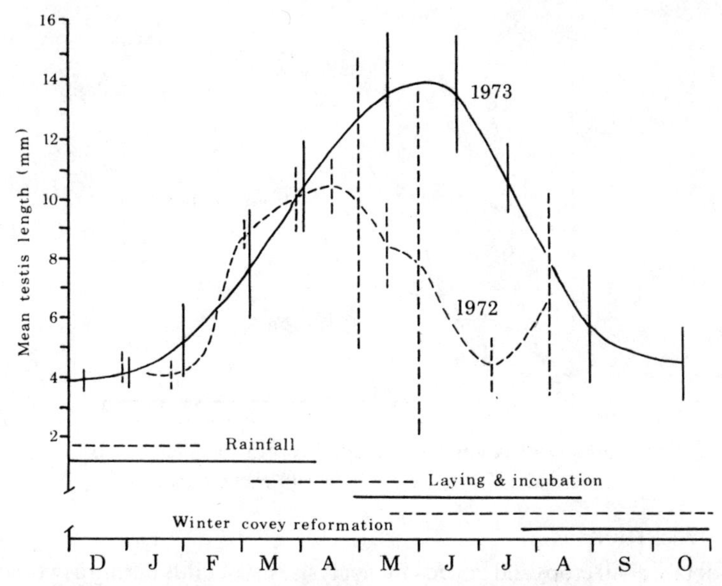

Figure 91. Recrudescence and regression of quail testes in 1972 and 1973, Shandon area. Bottom lines depict correlated events (dotted—1972; solid —1973).

somewhat earlier in 1972. But in 1972 testis size peaked at 10.4 mm in mid-April and dropped below 8.0 mm by the end of May. The effective breeding season, therefore, was 13 weeks. In 1973, on the other hand, testes continued to enlarge until early June, reaching a much higher average size (13.5 mm), and continued in breeding condition until early August, for a total fertile period of 20 weeks. Whereas the 1973 breeding season was somewhat delayed by cold rainy weather in February and March, the total effective period of breeding was much longer.

OVARIAN CYCLES—1972 AND 1973 COMPARED

Lewin (1963) and Anthony (1970a) used the criterion of ovarian weight to follow the reproductive cycle of female California Quail. My data showed similar development of ovaries in February and March of the two study years, but substantial differences thereafter (see Fig. 92). In 1972, ovaries developed rapidly in March, and four of five females collected late in that month had begun to lay. In one female, the ovary had seven ruptured follicles. A second had a well-developed incubation patch. However, laying terminated shortly thereafter. Of three females collected in mid-April, two had ceased laying and one of these had molted two primaries. By early May, ovarian weights had dropped to about the weights of early March, and egg laying was essentially over. In 1972, the length of the laying season was about six weeks.

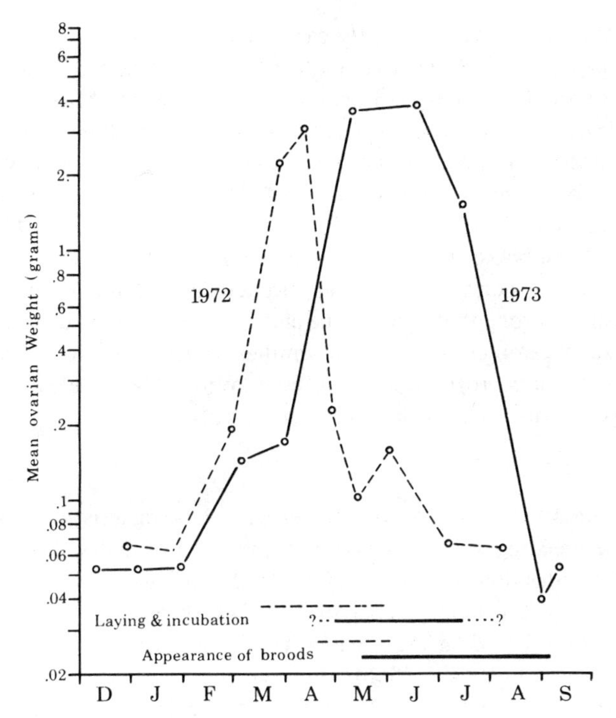

Figure 92. Recrudescence and regression of quail ovaries in 1972 and 1973.

In 1973, ovarian development slowed in March during the period of inclement weather, but accelerated in April and May. Laying began in mid-April and continued into mid-July, when two of five females collected were actively laying. Sightings of many young broods throughout August and a few in early September indicated that some birds continued laying until the end of July. In 1973, the laying season was approximately 13 weeks.

I collected data on weights of oviducts, and these reflected the same differences in timing and duration of the 1972 and 1973 breeding seasons as shown in ovary weights. For more complete details, reference is made to the original thesis (Erwin, 1975).

In summary, as evidenced by measurements of testes and by weights of ovaries and oviducts, the 1972 reproductive season in quail of the Shandon area started early but terminated quickly. In 1973, development of the gonads was retarded somewhat by inclement weather, but effective reproduction continued throughout the summer.

COVEY DISSOLUTION AND PAIRING

Behavioral differences among the quail paralleled the morphological differences in gonads. In late February of both years, I began seeing mated pairs within the winter coveys. By early March, the coveys were breaking up into small groups of four to eight, constituting clusters of pairs. Differences between 1972 and 1973 became apparent in late March. In 1972, pairing reached a peak in early April, but shortly thereafter, in late April, the birds began to regroup into coveys. By mid-May, pairs were few, groups of 20 to 30 birds were common, and I saw a group of about 100 foraging in the "home covey" area. On 14 May, about 350 birds roosted together there. For most birds, the breeding season was finished by late May 1972.

In 1973, by contrast, pair bonds strengthened during April and May, and continued strong throughout June and July. I saw no groups during this period except families or groups of families. After July, the rate at which grouping of adults progressed was difficult to judge because early-hatched young were indistinguishable from their parents.

"COW" CALLS

In the California Quail, unmated males give a characteristic "cow" call during the breeding season, presumably in search of mates. My field evidence from Shandon and observations by McMillan (1964) and Francis (1970) suggest that the frequency of "cow" calls is closely related to the intensity of the breeding urge, which in turn influences nesting success. In both years of my study, McMillan heard the first "cow" calls in early to mid-February during the early stages of testicular recrudescence, and I commonly heard "cow" calls in early March and throughout April. But in 1972 the calls ceased in early May, whereas in 1973 calling did not reach a maximum until May and did not subside until mid-July.

Although I recorded actual rates of "cow" calling in the home covey on several dates each year, I obtained satisfactory data for comparison between years only on 17 April 1972 and 1 April 1973. On each date the weather was similar, with fair skies and morning temperatures of 36°F (2.2°C) in 1972 and 30°F (−1.1°C) in 1973. On these dates, mean testicular lengths were similar (10.4 mm in 1972, 10.6 mm in 1973). I started counting at the first "cow" call in the early morning. On the 1972 date, between 0455 and 0642 hours, the average rate was 8.3 calls per minute, whereas in 1973, between 0539 and 0721 hours, the average rate was 21.5 calls per minute. On the first date, the quail ceased calling at 0642 hours, while on the second they continued calling all morning, but at a declining rate. In spite of the fact that there were about 40 percent fewer birds in the home covey the second year, the calling rate was 2.6 times faster than in 1972.

It is of particular interest to note that testis size was nearly identical on these two dates, suggesting that the breeding drive in male quail, as evidenced by "cow" calling, is not a simple function of testis development. I cannot guess what hormonal controls determine calling behavior, but the rate and persistence of "cow" calling is apparently a good predictive index of reproductive vigor and ultimate nesting success.

APPEARANCE OF BROODS

The first brood of young was seen on 21 April in 1972, and on 15 May in 1973 (McMillan field notes). The later sightings of chicks in 1973 corresponded with the later growth of ovaries in that year. In 1972, young broods were seen throughout May, but rarely thereafter. During the summer, I saw no broods that had hatched after the first week of June.

In 1973, early nesting was successful. Of 67 single or multiple broods that I observed between 27 May and 30 September, 30 had hatched prior to the second week of June, six prior to 15 May. New broods appeared throughout June, July, and August. I noted two broods that had hatched in September, both during the first week. The short time period in which broods appeared during the year of below-average rainfall and the long period in the year of above-average rainfall parallel earlier observations by McMillan (1964).

CARE OF YOUNG

A primary difference in quail behavior between 1972 and 1973 concerned the manner in which broods were raised. During the short breeding season of 1972, virtually all broods were reared by the parental pair. In 1973, the early broods were mostly turned over to cocks for rearing, while the females re-nested and participated in the rearing of the late broods.

Specifically, in 1972, 15 of 18 broods seen were accompanied by one or more pairs of adults, the other three broods by one or two adult males. In 1973, between 14 and 17 June, 11 of 13 broods were accompanied only by adult males, two by parental pairs. From 17 June to 30 September, I saw 54 more broods, all but 5 which were accompanied by pairs or by mixed groups of adults including females.

In summary, in 1973, lone males accompanied most broods hatched prior to mid-June, and pairs accompanied later-hatched broods. McMillan (1964:708) reported similar conditions in 1952, a year of high reproduction, when he "estimated that at least 75 percent of all young quail [hatched] . . . prior to July 1 were reared by male adults exclusively." He again noted this behavior in 1958, another year of high reproduction. In a one-year study, Anthony (1970a) noted that males reared 29 percent of early-hatched broods but only 17 percent of later broods. Gullion (1956a) observed that single males raised young Gambel Quail in a year of high reproduction. A presumed advantage of this parental strategy is that care of young by the male of a pair or by an unmated cock frees the female to incubate a second brood.

DOUBLE CLUTCHES AND BROODS

Under favorable circumstances, a female California Quail can raise two broods in the same year. W. Welch (see McLean, 1930) reported that a wild one-legged female successfully raised two broods in each of two breeding seasons. Francis (1965a) observed that, in a 100-foot square enclosure, two of three females hatched two successive broods. Anthony (1970a) reported taking two females that each had an incubation patch and reproductive tract in "laying condition" while accompanied by a mate and brood.

What appears to happen is that the female incubates and hatches the first clutch, helps attend it for about two weeks, then relegates the care of the young to her mate or to an unmated cock while she goes about laying a second clutch. This system would account for the large numbers of young, sometimes over 100, tended by adult males (Table 38). In 1973, I observed groups of 22 to 50 young that were tended by one adult male, and groups of over 100 tended by several adult males. Often these large troops included young of different sizes.

In the drier portions of its range, I believe, the California Quail augments its rate of reproduction during years of abundant moisture by females incubating two clutches of eggs. Two factors often cited as making it difficult for gallinaceous birds to successfully raise two broods are the short fertility period of the birds and the difficulty of raising the second brood to maturity during autumn weather. For California Quail in moist west-central California, Lewin (1963:266) examined the length of three laying seasons

TABLE 38.
Size and age of groups of young California Quail and Gambel Quail attended by adult males

Observer	*Date*	*Number of attendant males*	*Number of young*	*Age of young*
McMillan (1964)	6/21/52	1	> 50	unknown
	7/12/52	> 1	>100	unknown
	7/19/58	4	22	unknown
	7/19/58	1	19	unknown
	7/19/58	1	24	≈4 weeks
	7/20/58	8	167	>4 weeks
Gullion (1956a)	6/28/52	1	30	5–8 weeks
Erwin (present study)	6/12/71	1	≈ 30	6 weeks
	7/22/71	3	42	5 weeks
	6/15/73	1	36	3 and 5 weeks
	6/16/73	1	≈ 20	5 weeks
	6/17/73	1	22	5 weeks
	6/17/73	1	≈ 25	4 weeks
	6/19/73	1	49	5 weeks
	7/13/73	2	96	8–9 weeks
	7/17/73	>10	≈400	5–7 weeks

(19, 30, and 45 days) and concluded: "Second broods in a single season must be extremely rare and the data presented on laying periods for the three years in question in the Berkeley Hills show no indication of double broods." This conclusion seems inapplicable to the Shandon study area. Observations by McMillan (1964), Anthony (1970a), and myself suggest that in semi-arid areas double broods can and do occur in years of good rainfall and high reproduction. All three studies have noted long laying seasons and early broods raised by lone males. The observations of Gullion (1956a) and Raitt and Ohmart (1966) suggest double brooding in Gambel Quail. Possibly, quail in humid areas have the same capability but do not normally exhibit it.

Annual production of more than one brood per pair would be a significant adaptive advantage to the quail inhabiting the more arid environments. This capability would enable quail coveys that had declined to low numbers during several successive dry years to reproduce prodigiously in a favorable year. A few adult pairs that survived as seed stock could produce enough young to assure a carry-over of breeding stock through ensuing dry years.

PLUMAGE MOLT

The postnuptial molt of California Quail has been described by Genelly (1955) and Raitt (1961). Both authors noted that males begin molting one month earlier than females, but both sexes finish at about the same time. Molt ordinarily follows cessation of breeding, thus minimizing concurrent physiologic drains on the bird. I recorded the progress of molt among the birds collected during 1972 and 1973, and found the timing of the molting cycle to be very different in the two years.

To follow the progress of the postnuptial molt, I used Raitt's (1961) method of recording the sequential loss of the 10 primaries; for each feather, four growth substages were obtained by dividing the average length of the respective mature primary by four. For each specimen, I measured the newest primary and assigned it to a substage.

The molt data for individual males and females in 1972 and 1973 are shown in Figure 93. The molt analyses include all specimens taken after the commencement of molt, whether actually molting or not.

Between years, I found no significant difference, for either sex, in the rates of molt, but the timing of the molt was significantly different (P values $< .001$). In 1972, the first molting birds were collected the second week of April, but in 1973, not until the third week of June. Using Raitt's (1961) data on the rate of feather replacement, I backdated all molting birds to the start of their molt. In 1972, molt commenced the first week of April, whereas in 1973 it began the first week of June. Thus, in 1972, the year of low reproductive success, the molt began two months earlier. Also notable is the fact that in 1972 the earliest molting females were all immatures of the previous year (i.e., one year old). Earlier molting of immature females probably reflects an earlier cessation of the reproductive drive. Genelly

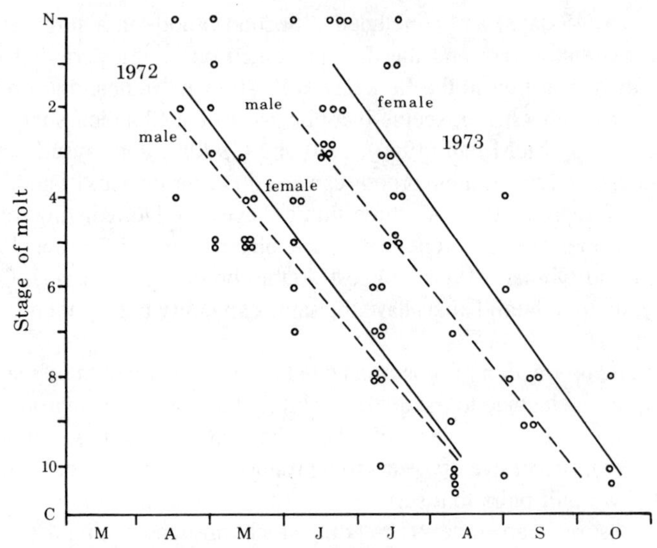

Figure 93. Sequence of primary molt in male and female quail at Shandon, in 1972 and 1973.

(1955) and Francis (1970) noted the lower breeding efficiency of immature female California Quail, and Lehmann (1953), studying the Texas bobwhite, witnessed lower reproductive success among immature females on ranges in dry condition.

In any event, the early cessation of breeding in 1972 was followed by early plumage molt. In 1973, when breeding continued throughout the summer, some birds started molting while still actively nesting. I collected two molting females that had eggs in their oviducts.

AGE RATIOS—SHANDON AREA

Reproductive success of the quail in 1972 and 1973 is reflected in the age ratios of birds shot in the Shandon area by McMillan and his hunting parties in the two years (extracted from Table 8):

Year	*Adult males*	*Adult females*	*Immature males*	*Immature females*	*Totals*	*Immature per 100 adults*
1972	86	67	14	25	192	25:100
1973	90	61	247	244	642	325:100

The enormous disparity in proportion of young quail in the bag between the two years (25:100 in 1972, 325:100 in 1973) is evidence enough of the difference in productivity.

AGE RATIOS THROUGHOUT CALIFORNIA

For comparison with the Shandon data, I collected information on sex and age of birds shot elsewhere in the state. Quail wings were collected for me at California Department of Fish and Game check stations and by cooperating sportsmen elsewhere. The wings were deposited in envelopes, one for each sex, and mailed to me at the close of the hunting season. Printed on the envelopes were instructions to take only the right wing from each bird, placing it in the envelope of the proper sex. During three shooting seasons (1971, 1972, and 1973), cooperators sent me a total of 14,388 wings. The sex and age distribution of these birds is shown in Table 39.

In the state as a whole, the age ratio in 1972 was 100 immature: 100 adults, whereas in 1973 the ratio was 282: 100. The aggregate sample suggests that 1973 was a superior year for quail production over a wide area.

However, when the data are plotted regionally, some marked differences are shown between mesic northern California and more arid southern California. Figures 94 and 95 depict the age ratios of individual samples distributed throughout the state.

In 1972, production of young was generally good in northern California but poor in the more arid southerly localities. In point of fact, I received relatively few samples from southern California because the scarcity of quail discouraged hunting.

In 1973, on the contrary, production in the north was moderate to poor, whereas in the south the ratio of young birds was very high, and I received many large samples of wings.

I suggest the following explanation. In 1972, the lower-than-average precipitation in the state was sufficient for quail to reproduce moderately well in northern California but not in the south. When rainfall increased in 1973, conditions were good for high reproduction and survival in the south, whereas the additional rainfall in the north, which normally receives two to four times the precipitation of southern California, was detrimental to production. Savage (1974) has documented the adverse effect of high rainfall on quail reproduction in Modoc County.

CENSUS OF QUAIL POPULATIONS

As a further check on fluctuations in quail populations, I counted birds on five areas in San Luis Obispo County during August and September in 1972 and 1973. The census areas were selected along a climatic gradient from the coast to the more arid interior. To count quail, I walked alone through suitable habitat, especially near water sources, locating coveys by

TABLE 39.
Statewide sex and age ratios of California Quail
obtained during three hunting seasons in California

Year	*Ad. ♂*	*Ad. ♀*	*Imm. ♂*	*Imm. ♀*	*Imm./100 Ad.*
1971	785	603	1435	1314	198/100
1972	986	686	840	845	100/100
1973	1048	735	2541	2470	282/100

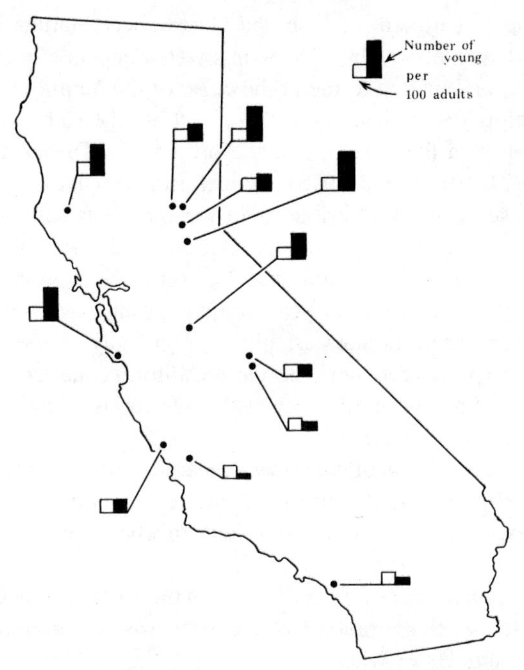

Figure 94. Age ratios of quail at various points in California in 1972.

tracks, calls, or direct observation. Not all birds were located, but the counts were made in the same manner and on the same areas both years, so the figures probably give a reliable index of abundance.

As shown in Table 40, at 9 of the 12 counting sites there was at least a doubling of quail numbers between 1972 and 1973. The rate of increase was fully as high toward the more mesic coast as in the arid interior. I rather expected a higher increase in the arid ranges. But age ratios of quail from the coast and interior (Figs. 94 and 95) confirmed that high production was widespread in 1973 south of San Francisco Bay, along the coast as well as inland.

DISCUSSION

What biological mechanism links the reproductive success of California Quail in their more arid habitats to annual rainfall? The key must lie in the realm of food availability or food nutritive value, which could affect the quail through physiological or psychological pathways, or both. Clearly, the annual differences in habitat which relate to rainfall must in some way alter the flow of pituitary and gonadal hormones that regulate reproduction. How might this mechanism function?

I foresee possible effects on quail both by the differing availability of food in a wet versus a dry year and by differing chemical composition of the food consumed.

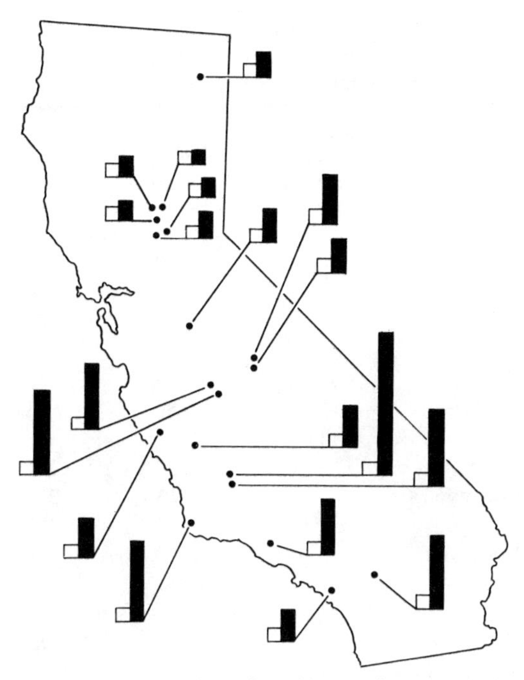

Figure 95. Age ratios of quail at various points in California in 1973.

The winter and spring diets of quail at Shandon differed substantially in 1972 and 1973, as shown in Table 11, Chapter 9, of this volume. Green leafage was available to the birds both years but was consumed sparingly in 1972 (26.7% volume of food in winter) and voraciously in 1973 (72.6% volume). Seeds constituted the bulk of the diet in spring 1972 but only one-quarter to one-half in 1973. Perhaps the difference in food habits between years related to differential availability of seeds, not greens.

Most seeds eaten by quail in the winter of 1971–72 were not produced in 1971 but were seeds that had lain dormant in the duff from a previous year, probably 1970. In 1971, few seeds of annual forbs, especially of legumes, were produced because severe aphid damage followed by heavy frosts injured or killed many of the annual plants. Nevertheless, seeds were available to the quail in the winter of 1971–72 because the low rainfall failed to germinate the combined earlier seed crops. Seed production was limited again in the spring of 1972 by drought. Thus, there were seed crops from one good year (1970) and two successive "poor" years available to quail in late fall of 1972. The low consumption of seeds in winter 1972–73 could be interpreted as lack of seed availability rather than preference for greens. Abundant rain that winter led to germination of most of the available seeds in the duff layer, which would further reduce the supply of seeds in 1973.

TABLE 40.
Post-breeding counts of California Quail from five localities in San Luis Obispo County

Counting area site	*Mean annual rainfall (inches)*	*1972 Populations*	*1973 Populations*	*% Increase*
Temblor Range	6			
Atriplex Guzzler		37	87	135
Phantom Guzzler		22	37	68
Tandem Guzzler		48	200	317
Spring Ranch	13			
Manzanita Canyon		136	283	108
Ranch House Covey		86	93	8
Camatta Ranch	11			
Camatta Canyon		557	1201	116
Little Navajo Canyon		115	325	183
Long Canyon		95	197	107
Sand Corrals Canyon		70	124	77
San Juan Creek		567	1453	156
Suey Ranch	15			
Cactus Hill Drainage		61	182	198
Montana de Oro State Park	16			
Coon Creek Covey		30	94	213
Total		1824	4276	$\bar{x}$ = 134

Availability of seeds may well dictate what the birds are eating but not how they respond physiologically to the diet. The great reproductive vigor shown by the birds in spring 1973, while they consumed roughly 50% seeds and 50% greens, requires additional explanation. Perhaps of more significance than seed quantity is the nutritive quality of the seed consumed. Legume seeds are notably more nutritious than the seeds of annual grasses, and there was a marked increase in the availability and the consumption of legume seeds in 1973 as compared to 1972. Table 22 summarizes the summer food habits of quail in the study area in the two years. In 1972, legume seeds constituted only 11.6% of the diet, whereas in 1973 they increased to 29.8%. Cultivated grains and other grass seeds decreased correspondingly from 51.3% in 1972 to 3.5% in 1973. Although these figures summarize samples collected in summer months (June, July, and August), the new crop of legume seeds began to ripen and be available to quail in April and May 1973. Nesting quail may be stimulated to continue vigorous nesting efforts by the easy availability of the new crop of legume seed.

Another factor that may influence nesting vigor is the chemical composition of the green foods. As noted in Chapter 9, chemical analysis of quail crop contents collected during these two years showed persistent occurrence of certain isoflavones (formononetin and genistein) in the foods consumed in spring 1972 and virtual absence of these compounds in the contents of crops collected in 1973. Feeding experiments which I conducted at Berkeley and which John Oh conducted at Hopland both showed that small quantities of these isoflavones, when added to normal diets of penned birds, substantially delayed sexual development and greatly reduced the number of eggs produced (see Leopold et al., 1976). The stunted forbs produced in spring 1972 apparently contained isoflavones in the leaves, whereas the vigorously growing plants of the same species growing under good moisture conditions in spring 1973 did not contain these inhibitory chemicals, at least after good growth started in February. This might be one mechanism that "turns off" quail breeding in a dry spring, and thus precludes the great physiologic expenditure of breeding in a year when the seed crop will be inadequate to rear young chicks or to support a large population in the subsequent winter.

There may well be other qualitative differences in diet between wet and dry years of which we have no knowledge. For example, the consumption of insects in 1973 was high through the spring and summer, which would add substantially to the protein intake of the birds. Insects were taken in less quantity in spring of 1972 and were virtually absent from the summer diet that year. The newly produced seeds of all types that began falling to the ground abundantly in April of 1973 might have had different qualitative values than the weathered seeds lying in the duff that supplied much of the quail diet in 1972.

I conclude merely that a wet spring at Shandon, such as 1973, produces food that may be both more abundant and more nutritious than that available in a dry year like 1972, and that in some complex manner this induces a profound physiologic effect on the quail and their reproductive behavior. For my part, I am inclined to stress the availability of legume seeds as a primary factor that stimulates prodigious production in a wet year. Isoflavones in greens may be one negative feedback mechanism that dampens reproductive effort in a dry year.

BIBLIOGRAPHY

Aginsky, B. W.

1943 Culture element distributions: XXIV, Central Sierra. Univ. Calif. Anthropol. Rec., 8:393–468.

American Ornithologists' Union

1957 Check-list of North American Birds. 5th ed. Baltimore Press, Baltimore. 691 pp.

Anthony, R. G.

1970a Ecology and reproduction of California Quail in southeastern Washington. Condor, 72:276–287.

1970b Food habits of California Quail in southeastern Washington during the breeding season. J. Wildl. Mgmt., 34:950–953.

Barclay, H. J., and A. T. Bergerud

1975 Demography and behavioral ecology of California Quail on Vancouver Island. Condor, 77:315–323.

Barnett, H. G.

1937 Culture element distributions: VII, Oregon Coast. Univ. Calif. Anthropol. Rec., 1:155–203.

Barrett, S. A.

1908 Pomo Indian basketry. Univ. Calif. Publ. Amer. Archaeol. Ethnol., 7:133–309.

1952 Material aspects of Pomo culture. I. Bull. Milwaukee Publ. Mus., 20:1–200.

Barrett, S. A., and E. W. Gifford

1933 Miwok material culture. Bull. Milwaukee Publ. Mus., 2:177–376.

Beals, R. L.

1933 Ethnology of the Nisenan. Univ. Calif. Publ. Amer. Archaeol. Ethnol., 31:335–414.

Beatley, J. C.

1969 Dependence of desert rodents on winter annuals and precipitation. Ecology, 50:721–724.

Beck, A.B.

1964 The oestrogenic isoflavones of subterranean clover. Austr. J. Agric. Res., 15:223–230.

Belding, L.

1892 Some of the methods and implements by which the Pacific Coast Indians obtained game. Zoe, 3:120–124.

Bennetts, H. W., C. J. Underwood, and F. L. Shier

1946 A specific breeding problem of sheep on subterranean clover pastures in Western Australia. Austr. Vet. J., 22:2–12.

Biely, J., and W. D. Kitts

1964 The anti-estrogenic activity of certain legumes and grasses. Can. J. Animal Sci., 44:297–302.

Biswell, H. H., R. D. Taber, D. W. Hedrick, and A. M. Schultz
1952 Management of chamise brushlands for game in the north coast region of California. Calif. Fish & Game, 38:453–484.

Bodenheimer, F. S., and F. Sulman
1946 The estrous cycle of *Microtus guentheri* D. and A. and its ecological implications. Ecology, 27:255–256.

Bolton, H. E.
1927 Fray Juan Crespi: missionary explorer on the Pacific Coast, 1769–1774. Univ. Calif. Press, Berkeley. 402 pp.

Braithwaite, L. W.
(in press) Breeding seasons of waterfowl in Australia. C.S.I.R.O. Wildl. Res., 14:73–78.

Braithwaite, L. W., and H. J. Frith
1969 Waterfowl in an inland swamp in New South Wales. III. Breeding. C.S.I.R.O. Wildl. Res., 14:65–109.

Burcham, L. T.
1959 Planned burning as a management practice for California wild lands. Proc. Soc. Amer. For., pp. 180–185.

California Dept. Fish & Game
1965 California's habitat types (8. Woodland chaparral and 9. Woodland grass), pp. 20–21. Calif. Fish & Game Plan, Vol. III, Support Data, Part A.

California Division of Mines and Geology
1959 Geological map of California. Olaf P. Jenkins edition, San Luis Obispo sheet and explanatory data. 1 map. 4 pp.

Campbell, H., and L. Lee
1956 Notes on the sex ratio of Gambel's and Scaled Quail in New Mexico. J. Wildl. Mgmt., 20:93–94.

Campbell, H., D. K. Martin, P. E. Fukovich, and B. K. Harris
1973 Effects of hunting and some other environmental factors on Scaled Quail in New Mexico. Wildl. Monogr., 34. 49 pp.

Carpenter, E. J., and R. E. Storie
1933 Soil survey of the Paso Robles area, California. U.S.D.A. Bur. Chem. and Soils. Series 1928, No. 34. 67 pp.

Chandler, R. E.
1970 Helminth parasites of California Quail (*Lophortyx californicus*) from the Okanagan Valley, British Columbia. Can. Jour. Zool., 48:741–744.

Clauson, S. G.
1959 Fire ant eradication—and quail. Alabama Cons., 30:14–15.

Cooper, W. S.
1922 The broad-sclerophyl vegetation of California: an ecological study of the chaparral and its related communities. Carnegie Inst., Washington, D.C. 124 pp.

Crispens, C. G., Jr., I. O. Buss, and C. F. Yocom
1960 Food habits of the California Quail in eastern Washington. Condor, 62:473–477.

Dana, S. T., and M. Krueger
1958 California lands: ownership, use, and management. Amer. For. Assn., Washington, D.C. 308 pp.

de Fremery, H.
1930 Valley quail and Sharp-shinned Hawk. Condor, 32:211.

De Shauensee, R. M.
1966 The species of birds of South America. Acad. Nat. Sci., Philadelphia. 577 pp.

Driver, H. E.
1936 Wappo ethnography. Univ. Calif. Publ. Amer. Archaeol. Ethnol., 36:179–220.
1937 Culture element distributions: VI, Southern Sierra Nevada. Univ. Calif. Anthropol. Rec., 1:53–154.

Drucker, P.
1937a Culture element distributions: V, Southern California. Univ. Calif. Anthropol. Rec., 1:1–52.
1937b The Tolowa and their southwest Oregon kin. Univ. Calif. Publ. Amer. Archaeol. Ethnol., 36:221–300.

Dutson, V. J.
1973 Use of the Himalayan blackberry, *Rubus discolor,* by the roof rat, *Rattus rattus,* in California. Calif. Vector News, 20:59–68.

Emlen, J. T., Jr.
1939 Seasonal movements of a low-density valley quail population. J. Wildl. Mgmt., 3:118–130.
1940 Sex and age ratios in survival of the California Quail. J. Wildl. Mgmt., 4:92–99.
1952 Flocking behavior in birds. Auk, 69:160–170.

Emlen, J. T., Jr., and B. Glading
1945 Increasing valley quail in California. Calif. Agr. Exp. Sta., Bull. 695. 56 pp.

Emlen, J. T., Jr., and F. W. Lorenz
1942 Pairing responses of free-living valley quail to sex-hormone pellet implants. Auk, 59:369–378.

Errington, P. L.
1934 Vulnerability of bob-white populations to predation. Ecology, 15: 110–127.

Errington, P. L., and F. N. Hamerstrom, Jr.
1935 Bob-white winter survival on experimentally shot and unshot areas. Iowa State Col. J. Sci., 9:625–639.
1936 The northern bob-white's winter territory. Iowa State Col. Agric. Res. Bull. 201, pp. 301–443.

Erwin, M. J.
1975 Comparison of the reproductive physiology, molt, and behavior of the California Quail in two years of differing rainfall. MS thesis in Wildland Resource Science, Univ. Calif., Berkeley. 72 pp.

Essene, F.
1942 Culture element distributions: XXI, Round Valley. Univ. Calif. Anthropol. Rec., 8:1–97.

Farmer, J. N., and R. P. Breitenbach
1966 A comparison of the course of *Plasmodium lophurae* infections in control and hormonally bursectomized chickens. Amer. Zool., 6: 307–308.

Faye, P. L.
1923 Notes on the Southern Maidu. Univ. Calif. Publ. Amer. Archaeol. Ethnol., 20:35–53.

Fitch, H. S., B. Glading, and V. House
1946 Observations on Cooper Hawk nesting and predation. Calif. Fish & Game, 32:144–154.

Fletcher, R. A.
1971 Effects of vitamin A deficiency on the pituitary-gonad axis of the California Quail, *Lophortyx californicus*. J. Exp. Zool., 176:25–34.

Foster, G. M.
1944 A summary of Yuki culture. Univ. Calif. Anthropol. Rec., 5:155–244.

Francis, W. J.
1965a Double broods in California Quail. Condor, 67:541–542.
1965b The effect of weather on population fluctuations in the California Quail (*Lophortyx californicus*). Ph.D. Dissertation. Univ. Calif., Berkeley.
1967 Prediction of California Quail populations from weather data. Condor, 69:405–410.
1970 The influence of weather on population fluctuations in California Quail. J. Wildl. Mgmt., 34:249–266.

Frith, H. J.
1959 The ecology of wild ducks in inland New South Wales. C.S.I.R.O. Wildl. Res., 4:97–181.
1967 Waterfowl in Australia. Angus and Robertson, Sydney. 328 pp.

Gallizioli, S.
1965 Quail research in Arizona. Arizona Game & Fish Dept., Phoenix. 12 pp.

Gassner, J. S.
1969 Voyages and adventures of La Perouse. Univ. Hawaii Press, Honolulu. 161 pp.

Gayton, A. H.
1948 Yokuts and western Mono ethnography, 1: Tulare Lake, Southern Valley, and Central Foothill Yokuts. Univ. Calif. Anthropol. Rec., 10:1–142.

Genelly, R. E.
1955 Annual cycle in a population of California Quail. Condor, 57:263–285.

Gifford, E. W., and A. L. Kroeber
1937 Culture element distributions: IV, Pomo. Univ. Calif. Publ. Amer. Archaeol. Ethnol., 37:117–254.

Glading, B.
1938a Studies on the nesting cycle of the California Valley Quail in 1937. Calif. Fish & Game, 24:318–340.

1938b A male California Quail hatches a brood. Condor, 40:261.

1943 A self-filling quail watering device. Calif. Fish & Game, 29:157–164.

1946 Upland game birds in relation to California agriculture. Trans. N. Amer. Wildl. Conf., 11:168–175.

Glading, B., H. H. Biswell, and C. F. Smith

1940 Studies on the food of the California Quail in 1937. J. Wildl. Mgmt., 4:128–144.

Glading, B., R. W. Enderlin, and H. A. Hjersman

1945 The Kettleman Hills quail project. Calif. Fish & Game, 31:139–156.

Glading, B., and R. W. Saarni

1944 Effect of hunting on a valley quail population. Calif. Fish & Game, 30:71–79.

Glading, B., D. M. Selleck, and F. T. Ross

1945 Valley quail under private management at the Dune Lakes Club. Calif. Fish & Game, 31:166–183.

Goddard, P. E.

1903 Life and culture of the Hupa. Univ. Calif. Publ. Amer. Archaeol. Ethnol., 1:1–88.

Goldschmidt, W.

1951 Nomlaki ethnography. Univ. Calif. Publ. Amer. Archaeol. Ethnol., 42:302–443.

Goodwin, D.

1953 Observations on voice and behavior of the Red-legged Partridge *Alectoris rufa*. Ibis, 95:581–614.

Gorsuch, D. M.

1934 Life history of the Gambel Quail in Arizona. Univ. Ariz. Biol. Sci. Bull. 2. 89 pp.

Gress, F.

1970 Reproductive status of the California Brown Pelican in 1970, with notes on breeding biology and natural history. Wildlife Branch Administrative Report No. 70-6. Calif. Dept. Fish & Game. 18 pp.

Grinnell, J.

1923 The burrowing rodents of California as agents in soil formation. J. Mamm., 4:137–149.

1927 A critical factor in the existence of southwestern game birds. Science, 65:528–529.

1932 Archibald Menzies, first collector of California birds. Condor, 34: 243–252.

Grinnell, J., H. C. Bryant, and T. I. Storer

1918 The game birds of California. Univ. Calif. Press, Berkeley. 642 pp.

Gullion, G. W.

1956a Evidence of double-brooding in Gambel Quail. Condor, 58:232–234.

1956b Let's go desert quail hunting. Nev. Fish & Game Comm. Biol. Bull. No. 2. 76 pp.

1960 The ecology of Gambel Quail in Nevada and the arid southwest. Ecology, 41:518–536.

Harper, H. T., B. H. Harry, and W. D. Bailey

1958 The Chukar Partridge in California. Calif. Fish & Game, 44:5–50.

Harrington, J. P.
1942 Culture element distributions: XIX, Central California coast. Univ. Calif. Anthropol. Rec., 7:1–46.

Heady, H. F.
1956 Examination and measurement of the California annual type. J. Range Mgmt., 9:25–27.

Heizer, R. F., editor
1972 George Gibbs' journal of Redick McKee's expedition through northwestern California in 1851. Archaeol. Res. Facil. Univ. Calif., Berkeley. 88 pp.

Henderson, C. W.
1971 Comparative temperature and moisture responses in Gambel and Scaled Quail. Condor, 73:430–436.

Herman, C.M.
1945 Gapeworm in California Quail and Chukar Partridge. Calif. Fish & Game, 31:68–72.

Herman, C. M., and A. I. Bischoff
1949 The duration of *Haemoproteus* infection in California Quail. Calif. Fish & Game, 35:293–299.

Herman, C. M., and B. Glading
1942 The protozoan blood parasite *Haemoproteus lophortyx* O'Roke in quail at the San Joaquin Experimental Range, California. Calif. Fish & Game, 28:150–153.

Herman, C. M., and H. Jankiewicz
1942 Reducing coccidiosis in California Valley Quail during captivity. Calif. Fish & Game, 28:148–149.

Hickey, J. J., editor
1969 Peregrine falcon populations: their biology and decline. Univ. Wisc. Press, Madison. 596 pp.

Howard, W. E., and J. T. Emlen, Jr.
1942 Intercovey social relationships in the Valley Quail. Wilson Bull., 54:162–170.

Hubbs, E. L.
1951 Food habits of feral house cats in the Sacramento Valley. Calif. Fish & Game, 37:177–189.

Hungerford, C. R.
1964 Vitamin A and productivity in Gambel's Quail. J. Wildl. Mgmt., 28:141–147.

Immelmann, K.
1971 Ecological aspects of periodic reproduction. In: Avian biology, Vol. 1. D. S. Farner and J. R. King, eds. Academic Press, N.Y., pp. 341–389.

Jenkins, D.
1957 The breeding of the Red-legged Partridge. Bird Study, 4:97–100.

Jepson, W. L.
1910 The silva of California. Univ. Calif. Press, Berkeley. 480 pp.
1930 The role of fire in relation to the differentiation of species in the chaparral. Fifth Internat. Bot. Congress, Cambridge Univ. Press, pp. 114–116.

Johnsgard, P. A.
1973 Grouse and Quails of North America. Univ. Nebraska Press, Lincoln. 553 pp.

Johnson, N. K.
1972 Origin and differentiation of the avifauna of the Channel Islands, California. Condor, 74:295–315.

Jones, R. E.
1968 The role of prolactin and gonadal steroids in the reproduction of California Quail, *Lophortyx californicus*. Ph.D. Thesis, Univ. Calif., Berkeley. 131 pp.

Keast, J. A., and A. J. Marshall
1954 The influence of drought and rainfall on reproduction in Australian desert birds. Proc. Zool. Soc. London, 124:493–499.

Kelly, I. T.
1932 Ethnography of the Surprise Valley Paiute. Univ. Calif. Publ. Amer. Archaeol. Ethnol., 31:67–210.

Kroeber, A. L.
1925 Handbook of the Indians of California. Bur. Amer. Ethnol. Bull. 78. Washington, D.C. 995 pp.
1929 The Valley Nisenan. Univ. Calif. Publ. Amer. Archaeol. Ethnol., 24:253–290.
1932 The Patwin and their neighbors. Univ. Calif. Publ. Archaeol. Ethnol., 29:253–423.

Kroeber, T.
1961 Ishi in two worlds. Univ. Calif. Press, Berkeley. 255 pp.

La Perouse, J. F. G. de
1798 A voyage round the world in the years 1785, 1786, 1787 and 1788. Ed. M. L. A. Milet-Mureau. 3 vols. London.

Lawrence, G. E.
1966 Ecology of vertebrate animals in relation to chaparral fire in the Sierra Nevada foothills. Ecology, 47:278–291.

Leach, H. R., and W. H. Frazier
1953 A study of the possible extent of predation on heavy concentrations of Valley Quail with special reference to the bobcat. Calif. Fish & Game, 39:527–538.

Lehmann, V. W.
1946 Bobwhite Quail reproduction in southwestern Texas. J. Wildl. Mgmt., 10:111–123.
1953 Bobwhite population fluctuations and vitamin A. Trans. N. Amer. Wildl. Conf., 18:199–246.

Leopold, A.
1931 Report on a game survey of the North Central States. Sporting Arms Ammun. Manufac. Inst., Madison, Wisc. 299 pp.
1933 Game management. Scribner's, N. Y. 481 pp.
1949 A sand county almanac. Oxford Univ. Press, N.Y. 226 pp.

Leopold, A. S.
1939 Age determination in quail. J. Wildl. Mgmt., 3:261–265.
1945 Sex and age ratios among Bobwhite Quail in northern Missouri. J. Wildl. Mgmt., 9:30–34.

1953 Intestinal morphology of gallinaceous birds in relation to food habits. J. Wildl. Mgmt., 17:197–203.

1954 The predator in wildlife management. Sierra Club Bull., 39:34–38.

1959 Wildlife of Mexico: the game birds and mammals. Univ. Calif. Press, Berkeley. 568 pp.

1972 The essence of hunting. National Wildlife, 10:38–40.

Leopold, A. S., M. Erwin, J. Oh, and B. Browning

1976 Phytoestrogens: adverse effects on reproduction in California Quail. Science, 191:98–100.

Lewin, V.

1963 Reproduction and development of young in a population of California Quail. Condor, 65:249–278.

1965 The introduction and present status of California Quail in the Okanagan Valley of British Columbia. Condor, 67:61–66.

Lewis, H. T.

1973 Patterns of Indian burning in California: ecology and ethnohistory. Ballena Press Anthropol. Pap., No. 1. 101 pp.

Liburd, E. M.

1969 Incidence of coccidia in California quail (*Lophortyx californicus*) from the Okanagan Valley, British Columbia. Can. Journ. Zool., 47:645–648.

Liburd, E. M., and J. L. Mahrt

1970 *Eimeria lophortygis* and *E. okanaganensis* n. sp. (Sporozoa: Eimeriidae) from California Quail *Lophortyx californicus* in British Columbia. Jour. Protozool., 17:252–253.

Linsdale, J. M.

1931 Facts concerning the use of thallium in California to poison rodents—its destructiveness to game birds and other valuable wild life. Condor, 33:92–106.

1936 The birds of Nevada. Pac. Coast Avifauna, 23:145 pp.

Loeb, E. M.

1926 Pomo folkways. Univ. Calif. Publ. Amer. Archaeol. Ethnol., 19: 149–404.

Loud, L. L., and M. R. Harrington

1929 Lovelock Cave. Univ. Calif. Publ. Amer. Archaeol. Ethnol., 25: 1–183.

Macgregor, W. G.

1950 The artificial roost—a new management tool for California Quail. Calif. Fish & Game, 36:316–319.

1953 An evaluation of California Quail management. Proc. 33rd Ann. Conf. Western Assn. State Game & Fish Comm., pp. 157–160.

Macgregor, W. G., Jr., and M. Inlay

1951 Observations on failure of Gambel Quail to breed. Calif. Fish & Game, 37:218–219.

Masson, W. Y., and R. U. Mace

1970 Upland game birds. Wildl. Bull. No. 5. Oregon State Game Comm. 44 pp.

Mayr, E.

1951 Speciation in birds. Intern. Ornith. Congr., 10:91–131.

McGilp, J. N
1924 Seasonal influences on the breeding of native birds. Emu, 24:155.

McLean, D. D.
1930 The quail of California. Calif. Game Bull., No. 2. 47 pp.

McMillan, I. I.
1960 Propagation of quail brush (Saltbush). Calif. Fish & Game, 46:507–509.
1964 Annual population changes in California Quail. J. Wildl. Mgmt., 28:702–711.

Merriam, C. H.
1967 Ethnographic notes on California Indian tribes. II. Ethnological notes on northern and southern California Indian tribes. Rep. Univ. Calif. Archaeol. Survey, No. 68, part II (Ed. R. F. Heizer), pp. 167–256.

Moss, R., A. Watson, and R. Parr
1974 A role of nutrition in the population dynamics of some game birds (Tetraonidae). Trans. 14th Congr. Internat. Union Game Biologists, pp. 193–201.
1975 Maternal nutrition and breeding success in red grouse (*Lagopus lagopus scoticus*). J. Anim. Ecol., 44:233–244.

Muller, C. H.
1966 The role of chemical inhibition (allelopathy) in vegetational composition. Bull. Torrey Bot. Club, 18:332–351.

Muller, C. H., and C. H. Chou
1972 Phytotoxins: an ecological phase of phytochemistry. In: Phytochemical ecology, ed. J. B. Harborne, Academic Press., N.Y., pp. 201–216.

Munro, J. A., and I. McT. Cowan
1947 A review of the bird fauna of British Columbia. Special Publ. No. 2., Prov. Mus., B. C.

Nielson, R. L.
1952 Factors affecting the California Quail populations in Uintah County, Utah. M.S. Thesis, Utah State Univ., Logan.

Nomland, G. A.
1938 Bear River ethnography. Univ. Calif. Anthropol. Rec., 2:91–126.

O'Roke, E. C.
1928 Parasites and parasitic diseases in the California Valley Quail. Calif. Fish & Game, 14:193–198.

Ortega y Gasset, J.
1972 Meditations on hunting. Scribner's, N.Y. 152 pp.

Post, G.
1951 Effects of toxaphene and chlordane on certain game birds. J. Wildl. Mgmt., 15:381–386.

Price, J. B.
1938 An incubating male California Quail. Condor, 40:87.

Rahm, N. M.
1938 Quail range extension in the San Bernardino National Forest—a progress report, 1937. Calif. Fish & Game, 24:133–158.

Raitt, R. J., Jr.
1960 Breeding behavior in a population of California Quail. Condor, 62: 284–292.

1961 Plumage development and molts of the California Quail. Condor, 63:294–303.

Raitt, R. J., and R. E. Genelly

1964 Dynamics of a population of California Quail. J. Wildl. Mgmt., 28: 127–141.

Raitt, R. J., and R. D. Ohmart

1966 Annual cycle of reproduction and molt in Gambel Quail of the Rio Grande Valley, southern New Mexico. Condor, 68:541–561.

Ralston, E. B.

1916 Shooting quail for market in San Mateo County. Calif. Fish & Game, 2:188.

Rawley, E. V., and W. J. Bailey

1972 Utah upland game birds. Utah State Div. Wildl. Res., Logan. 31 pp.

Reynolds, H. G.

1960 Life history notes on Merriam's kangaroo rat in southern Arizona. J. Mamm., 41:48–58.

Richardson, F.

1941 Results of the southern California quail banding program. Calif. Fish & Game, 27:234–249.

Robinson, T. S.

1957 The ecology of Bobwhites in south-central Kansas. U. Kans. Mus. Nat. Hist. & State Biol. Surv., Kans. Misc. Publ., No. 15. 84 pp.

Rosene, W.

1969 The Bobwhite Quail: its life and management. Rutgers Univ. Press, New Brunswick. 418 pp.

Rothermel, R. C., and C. W. Philpot

1973 Predicting changes in chaparral flammability. J. Forestry, 71:640–643.

Rudd, R. L., and R. E. Genelly

1956 Pesticides: their use and toxicity in relation to wildlife. Calif. Dept. Fish & Game, Game Bull. No. 7. 209 pp.

Rudkin, C. N. (Translator)

1959 The first French expedition to California, La Perouse in 1786. Glen Dawson, Los Angeles.

Sampson, A. W.

1944 Plant succession on burned chaparral lands in northern California. Calif. Agric. Exp. Sta. Bull., 685:1–144.

Sangler, H. M.

1931 Catalina cats and quail. Calif. Fish & Game, 17:450–541; 18:71 (1932).

Savage, A. E.

1974 Productivity and movement of California Valley Quail in northeast California, Trans. Western Sec. Wildl. Soc. Conf., pp. 84–88.

Sayama, K., and O. Brunetti

1952 The effects of sodium fluoroacetate (1080) on California Quail. Calif. Fish & Game, 38:295–299.

Schwartz, C. W., and E. R. Schwartz

1949 A reconnaissance of the game birds in Hawaii. Hawaii Bd. Comm. Agr. and Forestry, Fish and Game Div. 168 pp.

Serventy, D. L.
1971 Biology of desert birds. In: Avian biology, Vol. 1. D. S. Farner and J. R. King, eds. Academic Press, N.Y., pp. 287–339.

Shantz, H. L.
1947 Fire as a tool in management of brush ranges. Calif. Div. For., Sacramento. 156 pp.

Shaw, P.
1932 Studies on thallium poisoning in game birds. Calif. Fish & Game, 18:29–34.

Shields, P. W., and D. A. Duncan
1966 Fall and winter food of California Quail in dry years. Calif. Fish & Game, 52:275–282.

Smith, R. H., and S. Gallizioli
1965 Predicting hunter success by means of a spring call count of Gambel Quail. J. Wildl. Mgmt., 29:806–813.

Soumalainen, H., and E. Arhimo
1945 On the microbial decomposition of cellulose by wild gallinaceous birds (Family Tetraonidae). Ornis Fennica, 22:21–23.

Sparkman, P.
1908 The culture of the Luiseno Indians. Univ. Calif. Publ. Amer. Archaeol. Ethnol., 8:187–234.

Steward, J. H.
1933 Ethnography of the Owens Valley Paiute. Univ. Calif. Publ. Amer. Archaeol. Ethnol., 33:233–350.

Stewart, O. C.
1941 Culture element distributions: XIV. Northern Paiute. Univ. Calif. Anthropol. Rec., 4:361–446.
1963 Barriers to understanding the influence of use by fire by aborigines on vegetation. Tall Timbers Fire Ecol. Conf., 2:117–126.

Stoddard, H. L.
1931 The bobwhite quail: its habits, preservation and increase. Scribner's, N.Y. 559 pp.

Sumner, E. L., Jr.
1935 A life history study of the California Quail, with recommendations for its conservation and management. Calif. Fish & Game, 21:167–253, 275–342.

Swank, W. G., and S. Gallizioli
1954 The influence of hunting and of rainfall upon Gambel's Quail populations. Trans. N. Amer. Wildl. Conf., 19:283–297.

Tarshis, I. B.
1955 Transmission of *Haemoproteus lophortyx* O'Roke of the California Quail by hippoboscid flies of the species *Stilbometopa impressa* (Bigot) and *Lynchia hirsuta* Ferris. Exp. Parasit., 4:464–492.

Taverner, P. A.
1934 Birds of Canada. Nat. Mus. Canada Bull. 72. 445 pp.

Thornthwaite, C. W., and J. R. Mather
1957 Instructions and tables for computing potential evapotranspiration and the water balance. Drexel Inst. Tech., Lab. Climatology, Publ. Climat., 10:185–311.

True, G. H., Jr.
1933a The quail replenishment program. Calif. Fish & Game, 19:20–24.
1933b More about the quail refuges. Calif. Fish & Game, 19:248–252.
1934 An experiment in quail importation. Calif. Fish & Game, 20:365–370.
Twining, H.
1939 Some opinions of early California quail hunters. Calif. Fish & Game, 25:30–34.
Van De Graaff, K. M., and R. P. Balda
1973 Importance of green vegetation for reproduction in the kangaroo rat, *Dipodomys merriami merriami*. J. Mamm., 54:509–512.
Van Tyne, J., and A. J. Berger
1976 Fundamentals of ornithology. Wiley, N.Y. 808 pp.
Voegelin, E. W.
1938 Tubatulabal ethnography. Univ. Calif. Anthropol. Rec., 2:1–90.
1942 Culture element distributions: XX. Northeast California. Univ. Calif. Anthropol. Rec., 7:47–251.
Vogl, R. J., and P. K. Schorr
1972 Fire and manzanita chaparral in the San Jacinto Mountains, California. Ecology, 53:1179–1188.
Vorhies, C. T.
1928 Do southwestern quail require water? Amer. Nat., 62:446–452.
Weinmann, C. J., K. Murphy, J. R. Anderson, W. M. Longhurst, and G. Connolly
(In press) Epizootiological observations on the quail heartworm, *Splendidofilaria californiensis* (Weber and Herman, 1956) (Filarioidea). Canadian J. Zool.
Welch, W. R.
1928 Quail shooting in California today and fifty years ago. Calif. Fish & Game, 14:122–128.
1931a Game reminiscences of yesteryears. Calif. Fish & Game, 17:255–263.
1931b The propagation of California Quail in open breeding grounds. Calif. Fish and Game, 17:421–425.
Williams, G. R.
1952 The California Quail in New Zealand. J. Wildl. Mgmt., 16:460–483.
1957 Changes in sex ratio occurring with age in young California Quail in central Otago, New Zealand. Bird-Banding, 28:145–150.
1959 Ageing, growth-rate and breeding season phenology of wild populations of California Quail in New Zealand. Bird-Banding, 30:203–216.
Williams, H. W.
1969 Vocal behavior of adult California Quail. Auk, 86:631–659.

INDEX